*Basic Guide to
System Safety*

Basic Guide to System Safety

Third Edition

Jeffrey W. Vincoli
Manager of Compliance Assurance and Support Services
Bechtel Global Corporation

Published by John Wiley & Sons, Inc., Hoboken, New Jersey.
Published simultaneously in Canada.

For general information on our other products and services or for technical support, please contact our Customer Care Department within the United States at (800) 762-2974, outside the United States at (317) 572-3993 or fax (317) 572-4002.

Wiley also publishes its books in a variety of electronic formats. Some content that appears in print may not be available in electronic formats. For more information about Wiley products, visit our web site at www.wiley.com.

Library of Congress Cataloging-in-Publication Data:

Vincoli, Jeffrey W., author.
    Basic guide to system safety / Jeffrey W. Vincoli. – Third edition.
        p. ; cm.
    Includes index.
        ISBN 978-1-118-46020-7 (hardback)
    I. Title.
    [DNLM: 1. Occupational Health.   2. Safety.   3. Safety Management. WA 485]
    T55
    658.3′82–dc23   2013051270

Printed in the United States of America.

10 9 8 7 6 5 4 3 2

To my loving wife, Rosemary

Of all my accomplishments in this life, my greatest achievement was convincing you to be my wife. After more than 30 years together, I do not know how people go through life alone. I am blessed in many ways, but none more than having you as my wife. Thank you for always being there with your patience, your charm, your perspective, and your love. You are and will always be the most cherished thing about my life.

Contents

Preface

The third edition of the *Basic Guide to System Safety* contains all of the content of the previous editions, updated (where applicable) to reflect current industry practice. The first edition of the *Basic Guide to System Safety* was the first volume issued in a series of *Basic Guide* books that focused on the topics of interest to the practicing occupational safety and/or health professional. Other books in the Series include the *Basic Guide to Environmental Compliance, Basic Guide to Accident Investigation and Loss Control,* and *Basic Guide to Industrial Hygiene.* Each book has been designed to provide the reader with a fundamental understanding of the subject and attempt to foster a desire for additional information and training.

In addition to updated content of the previous editions, the revised third edition of the *Basic Guide to System Safety* introduces some system safety concepts not previously discussed to further expand upon the basic knowledge that is the cornerstone of the Basic Guide Series. In this regard, the third edition contains a discussion on the concept of Design for Safe Construction where the methods and techniques associated with the system safety discipline can be effectively utilized to identify, analyze, eliminate, or control system hazards during the design phase of a construction project. As with all analytical methods and techniques presented in this text, it is suggested that the concept of design for construction safety has definite application to general industry operations.

Also, information on the use of the various methods and techniques associated with the use of system safety has been expanded in the third edition to include guidance on the evaluation and verification of compliance efforts following the implementation of system safety analysis. This additional information will attempt to close-the-loop on the effective use of system safety analysis in the industrial safety environment.

It should be noted from the onset that it is not and never has been the intention of the *Basic Guide to System Safety* to provide any level of expertise beyond that of novice. Those practitioners and users who desire complete knowledge of the subject will not be satisfied with the information contained on these pages. It is not practical or feasible to expect a "basic guidebook" to contain all possible technical information on any subject, especially one as complex as system safety. However, those that require or perhaps only desire a basic understanding of a field similar but distinctly separate from their current area of specialization will find the third edition of *Basic Guide to System Safety* a valuable reference source and introductory primer. It is also assumed that those currently involved in the practice of system safety engineering and analysis might find this material somewhat enjoyable and, at the very least, refreshing. Also, professionals not directly involved in the system safety effort but who must work in association with those that are, will also find this text useful.

Finally, although the books in the *Basic Guide Series* were always originally intended for the practicing safety professional, the *Series* has been proven to be quite useful as textbooks for introductory courses in numerous colleges and universities. In this regard, the third edition will provide some additional fodder for enhancing existing primer courses on the subject.

It has long been known by practicing safety and health professionals that organizations with excellent safety performance records have a well-rounded corporate policy or at least a firmly established administrative posture that consistently emphasizes the importance and value of working safely. The leadership of such organizations has provided their strong (and intelligent) commitment in support of the safety effort. Therefore, this text concentrates especially upon the concepts that all executives should understand concerning the role that safety programs play in the successful operation of a business. No less of a commitment is necessary to properly implement system safety into an already established occupational/industrial safety and health program.

It is also recognized that, in order to achieve operationally safe system performance, system safety programs must be conducted with defined purpose, proficiency, skill, and a sense of well-rounded responsibility to the needs of the organization that the system safety program is intended to serve. In such a supportive environment, the system safety effort can and will become a vital contributor to the overall success of the enterprise.

This text places considerable emphasis on the integration of system safety principles and practices into the total framework of the organization. Anything less would constitute unsound business management. In the 20 years since the publication of the first edition of *Basic Guide to System Safety,* this very concept has been tested and proven viable numerous times by the author and other safety and health practitioners. There are examples of the successful integration of system safety methodologies into the practice of safety and health assurance in general industry, construction, rail, maritime, and aviation. It works, as long as there is understanding and commitment.

In short, the third edition of *Basic Guide to System Safety* follows tradition of the previous two editions. Safety and health professionals, as well as managers, engineers, technicians, designers, and college professors and their students should obtain some benefit from the information contained in this book.

ACKNOWLEDGMENTS

In the preparation of the third edition of *Basic Guide to System Safety*, I would like to thank and acknowledge those individuals and organizations that assisted in the initial, as well as revised, versions of this text.

First, I do not want to forget the valuable advice and assistance of those colleagues and associates who helped in the development and review of the first edition. Specifically, Steven S. Phillips, Frank Beckage, Douglas J. Tomlin, George S. Brunner, and Susie Adkins.

Second, I wish to recognize and acknowledge the training firm of Technical Analysis, Inc. (TAI) in Houston, Texas for permitting me to use some of their materials in the first and subsequent editions of this text, and for developing and providing exceptional training seminars on the subject of System Safety Engineering. Their contributions to the advancement of the System Safety discipline are commendable and appreciated.

Third, I would like to thank all those who participated in bringing this third edition of *Basic Guide to System Safety* to fruition including all the reference sources used herein, and the reviewers who helped identify specific areas for improvement over the previous editions. Thanks also to Fred Manuele for his leadership as Chair of the ANSI Z590.3-2011 Committee.

Fourth, a special thanks to Bob Esposito and Michael Leventhal of John Wiley & Sons for their support in making this third edition a reality.

Finally, I want to thank my wife, Rosemary, for her patience, understanding, and encouragement during my work to complete this process, and for her dedicated support of all that I do, always.

Part I

The System Safety Program

In the practice of occupational safety and health in industry today, the primary concern of any responsible organization is the identification and elimination of hazards that threaten the life and/or health of employees, as well as those which could cause damage to facilities, property, equipment, products, and/or the environment. When such risk of hazard cannot be totally eliminated, as is often the case, it becomes a fundamental function of the safety professional to provide recommendations to control those hazards in an effort to reduce the associated risk to the lowest acceptable levels.

It is the intention of this *Basic Guide to System Safety* to demonstrate the effectiveness of the system safety process in identifying and eliminating hazards, recommending risk reduction techniques, and methods for controlling residual hazard risk.

Part I will introduce the reader to the system safety process, how it evolved, how it can be managed, and how it relates to the current practice of the industrial safety and health professional. In fact, upon completion of Part I, the reader shall have developed a clear understanding of this relationship and, quite possibly, have developed an interest in the further pursuit of the system safety profession. As noted in the Preface, the information provided here is introductory in scope, intended to merely acquaint the reader with the system safety approach to hazard analysis and hazard risk reduction.

As a separate discipline, system safety had its origins in the aviation and aerospace industries. Systems safety has proven its worth in the dramatic improvements in

Basic Guide to System Safety, Third Edition. Jeffrey W. Vincoli.
© 2014 John Wiley & Sons, Inc. Published 2014 by John Wiley & Sons, Inc.

aviation safety over the past 60 years. It is not by chance that flying is demonstrably the safest mode of travel and this accomplishment has led to an undeniable understanding that all modern systems require a more logical, focused approach to identifying and controlling hazards. System safety is no longer a discipline reserved for the aerospace designer and nuclear engineer; it is the most effective method of improving the safety of any modern operation. As it has developed and matured, system safety has moved away from being the exclusive domain of design engineers and has become less mathematical or abstract and is now more practical and realistic. Modern concepts of system safety can be used by any organization or person who wants a logical, visible, and traceable method of identifying and controlling safety hazards and this is the objective of the *Basic Guide to System Safety.*

1

System Safety: An Overview

BACKGROUND

The idea or concept of *system safety* can be traced to the missile production industry of the late 1940s. It was further defined as a separate discipline by the late 1950s (Roland and Moriarty 1983) and early 1960s, used primarily by the missile, aviation, and aerospace communities. Prior to the 1940s, system designers and engineers relied predominantly on a trial-and-error method of achieving safe design. This approach was somewhat successful in an era when system complexity was relatively simple compared with those of subsequent development. For example, in the early days of the aviation industry, this process was often referred to as the *"fly-fix-fly"* approach to design problems (Roland and Moriarty 1983; Stephenson 1991) or, more accurately, *"safety-by-accident."* Simply stated, an aircraft was designed based upon existing or known technology. It was then flown until problems developed or, in the worst case, it crashed (Figure 1.1). If design errors were determined as the cause (as opposed to human, or "pilot" error), then the design problems would be fixed and the aircraft would fly again. Obviously, this method of after-the-fact design safety worked well when aircraft flew low and slow and were constructed of wood, wire, and cloth. However, as systems grew more complex and aircraft capabilities such as airspeed and maneuverability increased, so did the likelihood of devastating results from a failure of the system or one of its many subtle interfaces. This is clearly demonstrated in the early days of the aerospace era (the 1950s and 1960s). As the industry began to develop jet powered aircraft and space and missile systems, it quickly became clear that engineers

Basic Guide to System Safety, Third Edition. Jeffrey W. Vincoli.
© 2014 John Wiley & Sons, Inc. Published 2014 by John Wiley & Sons, Inc.

Figure 1.1 The "fly-fix-fly" approach, or more accurately "safety-by-accident," focused on fixing design issues after an accident event rather than focusing on accident prevention through design.

could no longer wait for problems to develop; they had to anticipate them and "fix" them before they occurred. To put it another way: the "fly-fix-fly" philosophy was no longer feasible. Elements such as these became the catalyst for the development of *systems engineering*, out of which eventually grew the concept of *system safety*. The need to anticipate and fix problems before they occurred led to a new approach—a consideration of the design as a "system." This means that all aspects of the design of operation (e.g., machine, operator, and environment) must be considered in identifying potential hazards and establishing appropriate controls. Another important part of this "systems" approach to safety is the realization that resources for safety are limited and there must be some logical, reasoned way to apply resources to the most serious potential problems. Systems safety provides this capability. Figure 1.2 shows a simplification of the basic elements of the systems engineering process. It is noted that safety comprises only one part of this integrated engineering design approach (Larson and Hann 1990). Taken one step further, Figure 1.3 demonstrates how the systems approach associated with the initial element of the systems safety engineering process—the design aspect—can support the identification of hazards in the earliest phases of a project life cycle. Only after the accurate identification of hazards can proper elimination or control measures be determined.

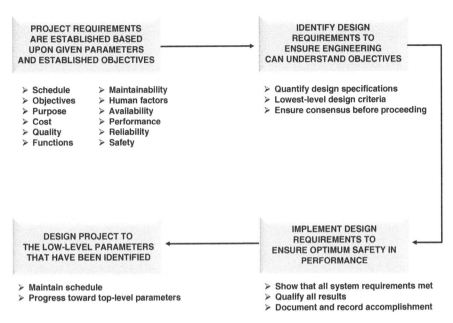

Figure 1.2 *The system safety engineering process* (Source: *Larson and Hann 1990*).

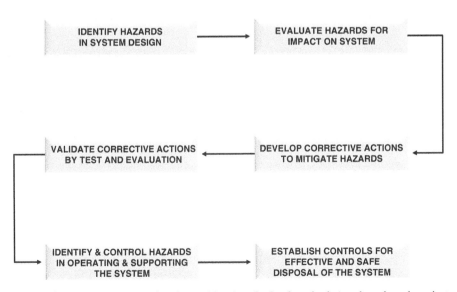

Figure 1.3 *The systems approach to the consideration of safety from the design phase through product disposal or project termination.*

The dawn of the manned spaceflight program in the mid-1950s also contributed to the growing necessity for safer system design. Hence, the growing missile and space systems programs became a driving force in the development of system safety engineering. Those systems under development in the 1950s and early 1960s required a new approach to controlling hazards such as those associated with weapon and space systems (e.g., explosive components and pyrotechnics, unstable propellant systems, and extremely sensitive electronics). The Minuteman Intercontinental Ballistic Missile (ICBM) was one of the first systems to have had a formal, disciplined, and defined system safety program (Roland and Moriarty 1983). In July of 1969, the US Department of Defense (DOD) formalized system safety requirements by publishing MIL-STD-882 entitled "System Safety Program Requirements." This Standard has since undergone a number of revisions.

The US National Aeronautics and Space Administration (NASA) soon recognized the need for system safety and has since made extensive system safety programs an integral part of space program activities. The early years of our nation's space launch programs are full of catastrophic and quite dramatic examples of failures. During those developing years, it was a known and quite often stated fact that *"our missiles and rockets just don't work, they blow up."* The many successes since those days can be credited in large part to the successful implementation and utilization of a comprehensive system safety program. However, it should be noted that the Challenger disaster in January 1986 and the loss of the orbiter Columbia upon reentry in February of 2003 stand as historic reminders to us all that, no matter how exact and comprehensive a design or operating safety program is considered to be, the proper *management* of that system is still one of the most important elements of success. This fundamental principle is true in any industry or discipline.

Eventually, the programs pioneered by the military and NASA were adopted by industry in such areas as nuclear power, refining, mass transportation, chemicals, healthcare, and computer programming.

Today, the system safety process is still used extensively by the various military organizations within the DOD, as well as by many other federal agencies in the United States such as NASA, the Federal Aviation Administration, and the Department of Energy. In most cases, it is a required element of primary concern in the federal agency contract acquisition process.

Although it would not be possible to fully discuss the basic elements of system safety without comment and reference to its military/federal connections, the primary focus of this text shall be placed upon the advantages of utilizing system safety concepts and techniques as they apply to the general safety arena. In fact, the industrial workplace can be viewed as a natural extension of the past growth experience of the system safety discipline. Many of the safety rules, regulations, statutes, and basic safety operating criteria practiced daily in industry today are, for the most part, the direct result of a real or perceived need for such control doctrine. The requirement for safety controls (written or physical) developed either because a failure occurred or someone with enough foresight anticipated a possible failure and implemented controls to avoid such an occurrence. Even though the former example is usually the case, the latter is also responsible for the development of countless safe operating requirements practiced in industry today. Both, however, are also the basis upon which system safety engineers operate.

The first method, creating safety rules *after* a failure or accident, is likened to the *"fly-fix-fly"* approach discussed earlier. The second method, anticipating a potential failure and attempting to avoid it with control procedures, regulations, and so on, is exactly what the system safety practitioner does when analyzing system design or an operating condition or method. However, when possible or practical, the system safety concept goes a step further and actually attempts to engineer the risk of hazard(s) out of the process. With the introduction of the system safety discipline, the *fly-fix-fly* approach to safe and reliable systems was transformed into the *"identify, analyze, and eliminate"* (Abendroth and Grass 1987) method of system safety assurance.

We have established the basic connection between the system safety discipline and its relationship to the general industry occupational safety practice. This conceptual relationship will be examined in more detail throughout this text.

THE DIFFERENCE BETWEEN INDUSTRIAL SAFETY AND SYSTEM SAFETY

Industrial safety, or occupational safety, has historically focused primarily on controlling injuries to employees on the job. The industrial safety engineer usually is dealing with a fixed manufacturing design and hazards that have existed for a long time, many of which are accepted as necessary for operations. Traditionally, more emphasis is often placed on training employees to work within this environment rather than on removing the hazards.

To perform their charter, industrial safety engineers collect data during the operational life of the system and eliminate or control unacceptable hazards where possible or practical. When accidents occur, they are investigated and action is taken to reduce the likelihood of a recurrence—either by changing the plant or by changing employee work rules and training. The hazards associated with high-energy or dangerous processes are usually controlled either by

- Disturbance control algorithms implemented by operators or an automated control system or
- Transferring the plant to a safe state using a separate protection system.

Safety reviews and compliance audits are conducted by industrial safety organizations within a company or, less frequently, by safety committees to ensure that unsafe conditions in the workplace are corrected and that employees are following the work rules specified in manuals, directives, and operating instructions. Lessons learned from accidents are incorporated into design standards, and much of the emphasis in the design of new plants and work rules is on implementing these standards. Often, the standards are enforced by the government through occupational safety and health legislation.

In contrast, system safety has been traditionally concerned primarily with new systems. The concept of "loss" is treated much more broadly as relevant losses may include

- Injury to nonemployees;
- Damage to equipment, property, or the environment;
- Loss of mission.

As has been previously established, instead of making changes as a result of operational experience with the system, system safety attempts to identify potential hazards before the system is designed, to define and incorporate safety design criteria, and to build safety into the design before the system becomes operational. Although standards are used in system safety, they usually are "process standards" rather than "product standards" as reliance on design or product standards is often inadequate for new types of systems, and more emphasis is placed on upfront analysis and designing for safety. There have been attempts to incorporate system safety techniques and approaches into traditional industrial safety programs, especially when new plants and processes are being built. Although system safety techniques are considered "overkill" for many industrial safety problems, larger organizations and increasingly dangerous processes have raised concern about injuries to people outside the workplace (e.g., pollution) and have therefore made system safety approaches more relevant. Furthermore, with the increase in size and cost of plant equipment, changes and retrofits to increase safety are costly and may require discontinuing operations for a period of time. Similarly, it is interesting to note that system safety is increasingly considering issues that have been traditionally thought to be strictly industrial safety concerns.

In summary, industrial safety activities are designed to protect workers in an industrial environment with extensive standards imposed by federal codes or regulations to provide for a safe workplace. However, with few exceptions, these codes seldom apply to protection of the product being manufactured. With the increasingly more frequent use of robotics in the workplace environment and with long-lived engineering programs like space launch vehicles that have substantial continuing complex engineering design activities, the traditional concerns of industrial safety and system safety have become more intertwined (Leveson 2005).

In 2011, these circumstances have led to the development of a new American National Standards Institute/American Society of Safety Engineers (ANSI/ASSE) Standard titled *Prevention Through Design: Guidelines for Addressing Occupational Hazards and Risks in Design and Redesign Processes* (ANSI/ASSE Z590.3-2011). This Standard and its relationship to the objectives of this Basic Guide to System Safety will be discussed further in Chapter 4.

SYSTEM SAFETY AND THE ASSESSMENT OF RISK

The idea, concept, or process of system safety has been defined in many ways, by a wide variety of scientific and technical professionals. However, since its inception, system safety has had the specific, driving purpose to eliminate system faults or failure risk and subsequent recognized accident and/or hazard potential through design and implementation of engineering controls. Basically, system safety can be defined as:

> *a sub-discipline of systems engineering that applies scientific, engineering and management principles to ensure adequate safety, the timely identification of hazard risk, and initiation of actions to prevent or control those hazards throughout the life cycle and within the constraints of operational effectiveness, time, and cost (Stephenson 1991).*

In the simplest of terms, system safety uses systems theory and systems engineering approaches to prevent foreseeable accident events and to minimize the result of unforeseen events. Losses in general (not just human death or injury) are considered and can include destruction of property, loss of mission, and/or environmental harm (Leveson 2005). The term *safety*, as used here, is somewhat relative. Although "safety" has often been traditionally defined in many sources as .... *freedom from those conditions that can cause death, injury, occupational illness, or damage to or loss of equipment or property* (MIL-STD-882), it is generally recognized in the profession that this definition is somewhat unrealistic (Leveson 1986). This definition would indicate that *any* system containing some degree of risk is considered *unsafe*. Obviously, this is not practical logic since almost any system that produces some level of personal, social, technological, scientific, or industrial benefit contains an indispensable element of risk (Browning 1980). For example, safety razors or safety matches are not entirely *safe*, only *safer* than their alternatives. They present an acceptable level of risk while preserving the benefits of the less-safe devices that they have replaced (Leveson 1986). A more vivid example of risk reduction and acceptance involves the sport of skydiving: Most sane skydivers would never jump out of an airplane without a parachute. The parachute provides a *control measure* intended to eliminate some level of risk. However, even with the parachute strapped in place, the jumper is still accepting the risk of parachute failure. System safety is concerned with the aspect of reducing the risk(s) associated with a hazard to its lowest acceptable level. In reality, no aircraft could fly, no automobile could move, and no ship could be put out to sea if *all* hazards and *all* risk had to be completely eliminated first (Hammer 1972). Similarly, no drill press could be operated, forklift driven, petroleum refined, dinner cooked, microwave used, water boiled, and so on, without some element of operating risk.

This problem is further complicated by the fact that attempts to eliminate risk result instead in the often unfortunate displacement of risk (Malasky 1982). For example, some approved (by the US Food and Drug Administration) preservatives currently utilized in the food processing industry to prevent bacteria growth and spoilage are, themselves, a suspected cause of cancer (e.g., sodium nitrates). Likewise, there is a risk trade-off between the known benefits of improved medical diagnosis and treatment which result from the use of radiation (e.g., X-rays, radiation therapy), against the known risks of human exposure to radiation. Hence, safety is really more of a relative issue in that nothing is *completely* safe under *all* circumstances or *all* conditions. There is always some example in which a relatively safe material or piece of equipment can become hazardous. The very act of drinking water, if done to excess, can cause severe renal problems in most cases (Gloss and Wardel 1984).

Unfortunately, the question *"How safe is safe enough?"* has no simple answer. For example, it is not uncommon to hear the term *"99.9% risk-free"* used to signify high assurance or low-risk assessments, especially in the advertising industry. In fact, it would be safe to say that this terminology is somewhat overused in our society. However, consider the following statistical facts (Larson and Hann 1990).

In the United States today, 99.9% safe would mean:

- one hour of unsafe drinking water per month;
- 20,000 children per year suffering from seizures or convulsions due to faulty whooping cough vaccinations;
- 16,000 pieces of mail lost per hour;
- 500 incorrect surgical operations each week;
- 50 newborns dropped by doctors each day.

Clearly, a 99.9% assurance level is not really "safe enough" in today's society. If the percentage were increase by a factor of ten to "99.99%," the following information indicates that this level of risk is still unacceptable in certain instances. A 99.99% risk-free assurance level would mean:

- 2000 incorrect drug prescriptions per year;
- 370,000 checks deducted from the wrong account per week;
- 3200 times per year, your heart would fail to beat;
- 5 children sustaining permanent brain damage per year because of faulty whooping cough vaccinations.

Obviously, the need to ensure optimum safety in a given system, industry, or process is absolutely essential. In fact, with certain critical functions of a system, there is no room for error or failure, as is evidenced in some of the above listed examples. Thus, *safety* becomes a function of the situation in which it is measured (Leveson 1986).

Therefore, the question still remains as to the proper definition of *safety*. One possible improvement of the previously presented MIL-STD-882 definition might be that safety .... *is a measure of the degree of freedom from risk in any environment* (Leveson 1986). Hence, *safety* in a given system or process is not measured so much as is the level of *risk* associated with the operation of that system or process. This fundamental concept of acceptable risk is the very foundation upon which system safety has developed and is practiced today.

In the world of occupational safety, the ever-present requirement to achieve 100% compliance with written codes, rules, regulations, or established operating procedures is a challenge in and of itself. However, in the practice of system safety, it must be clearly understood that *"design by code"* is no substitute for intelligent engineering and that codes only establish a minimum requirement which, in many systems or situations, must be exceeded to ensure adequate elimination or control of identified hazard(s). Therefore, 100% "compliance" usually means a system has met only the *minimum* safety requirements. Looking at the subject of regulatory compliance a different way, let us consider what it really means to be 100% compliant with the minimum requirements established by applicable codes and regulations. In the United States, for example, the Occupational Safety and Health Administration (OSHA) claims that occupational injuries and fatalities have decreased between 60% and 65% during the 40-year period of it existence between 1971 and 2011. While such a statistic is certainly laudable for obvious reasons, it also tells us that between 30% and 35% of workers in the United States are still suffering occupational injuries or fatalities. Clearly, compliance with the minimum requirements established by OSHA is not enough. Employers must do more. They must go beyond compliance, where required, to ensure optimum safety and health in the workplace.

The efforts associated with system safety attempt to exceed these minimum compliance standards and provide the highest level of safety (i.e., the lowest level of acceptable risk) achievable for a given system. In addition, it is important to mention at this point that system safety has often been used to demonstrate that some compliance requirements can be too excessive while providing insufficient risk reduction to justify the costs incurred. Costs, such as operating restrictions, system performance, operational schedules, downtime, and, of course, actual dollars, are all elements of a successful operation which must be considered when determining the validity of implementing any new compliance controls. Proper utilization of system safety engineering has proven to be an excellent tool for evaluating the value of such controls with regard to actual savings and reduction of risk. For example, in general, the OSHA requires that machine guarding be employed to protect operators of machines from hazards created by the machining point of operation and/or other hazards associated with machine operation [OSHA 29 CFR §1910.212(a)(1)]. Safety practitioners and machine operators both are well aware that a machine can be effectively guarded to the point where it is no longer usable and, in actuality, borders on the ridiculous. Safety professionals will recall the famed "OSHA Cowboy" which was first drawn by J. N. Devin in 1972 and has circulated throughout the industry ever since. As shown in Figure 1.4, the OSHA Cowboy was a satirical view of OSHA compliance extremes. Essentially, the cartoon drawing demonstrated that the risks to the cowboy on horseback can be guarded and controlled to the point where even simple movement would be impossible.

As stated previously, system safety developed or evolved as a direct result of a need to ensure, to the greatest extent possible, reliability in the safe operation of a system or set of systems (especially when a given system is known to be hazardous in nature). While no system can be considered completely or 100% reliable, system safety is an attempt to get as close as practical to this goal. Over the years, numerous techniques and methods used to formally accomplish the system safety task have also evolved and have further expanded our capabilities to examine systems, identify hazards, eliminate or control them, and reduce risk to an acceptable level in the

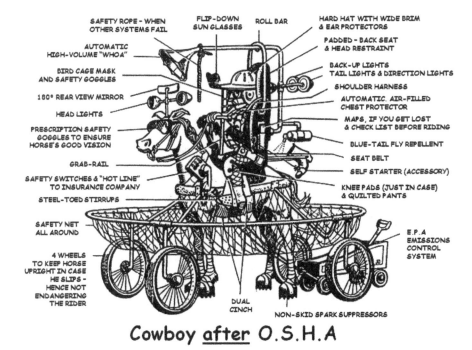

Cowboy after O.S.H.A

Figure 1.4   *The "OSHA Cowboy" as first depicted by J.N. Devin in 1972.*

operation of that system. These analytical methods and/or techniques are known by many names such as, but certainly not limited to the following common system safety tools:

- Preliminary Hazard Analysis (PHA)
- System Hazard Analysis (SHA)
- Subsystem Hazard Analysis (SSHA)
- Operating & Support Hazard Analysis (O&SHA)
- Failure Mode and Effect Analysis (FMEA)
- Fault Tree Analysis (FTA)
- Fault (or Functional) Hazard Analysis (FHA)
- Management Oversight and Risk Tree (MORT)
- Energy Trace and Barrier Analysis (ETBA)
- Sneak Circuit Analysis (SCA)
- Software Hazard Analysis (SWHA)
- Common Cause Failure Analysis (CCFA)
- Cause and Effect Analysis (CEA)
- Event Tree Analysis (ETA)

- Hazard and Operability Studies
- Random Number Simulation Analysis (RNSA)
- Health Hazard Analysis (HHA)

The chapters in Part II of this text will provide a simplified explanation of the most common used of these techniques. The intention is to present a *basic foundation of understanding* with regard to the fundamental analytic methods associated with the system safety engineering discipline. It is important to note once again that it is not the purpose of this limited volume to provide a single-source technical reference on the complete scope of the system safety discipline. This approach, although feasible, is not practical or advisable when attempting to discuss only the basics of system safety development and its potential use in general industry. There are numerous scientific and engineering reference volumes available on this subject and further research is recommended for those that desire more complete and detailed instruction on the use of system safety techniques. In addition, many universities, training institutions, professional and trade organizations, and independent private consultants offer continuing educational courses on the subject of system safety engineering/analysis.

2

System Safety Concepts

FUNDAMENTALS

Since its initial development a half-century ago, the system safety discipline has experienced a dramatic evolution of change and growth. Some analysts have compared this rapid development to the humorous analogy of a man that walked into a doctor's office with a frog growing from his forehead. When the doctor asked: *"How did it happen?"* The frog replied: *"It started as a pimple on my rear end!"* (Olson, undated).

Although, as defined in Chapter 1, system safety has emerged as a subdiscipline within systems engineering, it has quickly become an essential element of the safety planning process in many industries including nuclear, aerospace/aviation, refining, healthcare, and so on. In order to properly understand system safety as utilized in this text, a fundamental understanding of some basic safety concepts, principles, and terms must first be examined. The following definitions, from the Glossary of Terms, are therefore provided here for discussion purposes:

System: A combination of people, procedures, facility, and/or equipment all functioning within a given or specified working environment to accomplish a specific task or set of tasks (Stephenson 1991).

Safety: A measure of the degree of freedom from risk or conditions that can cause death, physical harm, or equipment/property damage (Leveson 1986). Note: assumption of *risk* is an essential ingredient of system safety philosophy.

System Safety Precedence: An ordered listing of preferred methods of eliminating or controlling hazards (MIL-STD-882).

Basic Guide to System Safety, Third Edition. Jeffrey W. Vincoli.
© 2014 John Wiley & Sons, Inc. Published 2014 by John Wiley & Sons, Inc.

Hazard: A condition or situation which exists within the working environment capable of causing harm, injury, and/or damage.

Hazard Severity: A categorical description of hazard level based upon real or perceived potential for causing harm, injury, and/or damage.

Hazard Probability: The likelihood that a condition or set of conditions will exist in a given situation or operating environment.

Mishap: An occurrence which results in injury, damage, or both.

Near-miss: An occurrence which could have resulted in injury, damage, or both, but did not.

Risk: The likelihood or possibility of hazard consequences in terms of severity and probability (Stephenson 1991).

THE SYSTEM SAFETY PROCESS

The process of system safety revolves around a desire to ensure that jobs or tasks are performed in the safest manner possible, free from unacceptable risk of harm or damage. The primary concern of system safety is the management of hazards: their identification, evaluation, elimination, and control through analysis, design, and management procedures (Leveson 2005). This forward-looking process occurs within a working environment where people, operating procedures, equipment/hardware, and facilities are all integral factors which may or may not affect the safe and successful completion of the job or task. Each of these elements themselves might also impose some degree of risk or hazard to people or equipment during the performance of a task. People, for example, can be hazardous to themselves or others in an industrial or technological working environment. Inattention, lack of proper or adequate training, horseplay, fatigue, stress as well as substance abuse, personal problems (marriage, financial, etc.) are all "human" factors that interfere with optimum or desirable human work performance. Likewise, certain equipment or tools can present hazards, even if operating as intended (pressure systems, nuclear reactors, powder-actuated hand tools, etc.). Also, inadequately written or faulty operating instructions and procedures can cause hazards to operational or task flow. Therefore, the system safety process must take each of these factors into consideration to properly address the variety of potential hazards that might be associated with a specific task or job. Figure 2.1 is a graphic representation of the system safety process which incorporates the concept of people, procedures, facility, and/or equipment that must operate within a specific work environment to accomplish a task or set of tasks (Stephenson 1991; Roland and Moriarty 1983). For example, consider a forklift operator involved in relocating several drums of a highly volatile, flammable solvent from one location of a plant to another. What potential or degree of risk exists for a failure or accident in a simple operation such as this? In answering this question, one should think about the operator, his/her training, and level of experience. The forklift and other associated equipment (drum handling attachment, securing devices, etc.) must also be evaluated

Figure 2.1   *The elements of the system safety process* (Source: *Stephenson 1991*).

as potential sources of operational failure. The facility in which the drums are located should be designed to store such commodities. Fire suppression equipment must be evaluated for adequacy. Normal operating procedures as well as emergency/spill control requirements should be examined for proper considerations/controls. This analysis of hazard or risk potential can become quite detailed. However, for the purpose of this illustration, the point of risk analysis of system or process operations should be obvious. As one can see by this simple example, there is a great deal of hazard potential associated with the above described task. It is the function of system safety to pursue such an evaluation to the greatest extent possible, with respect to the complexity of the task, system, operation, or procedure.

The system safety discipline will require the timely identification and subsequent evaluation of the hazards associated with this operation, *before* losses occur. The hazards must then be either eliminated or controlled to an *acceptable level of risk* in order to accomplish the goal of relocating the hazardous chemicals. In short, the system safety process will identify any corrective actions which must be implemented before the task is permitted to proceed. The *fly-fix-fly* approach discussed earlier has also been described as an *"after-the-fact"* attempt to improve operational safety performance. In contrast, the system safety concept requires *"before-the-fact"* control of system hazards.

SYSTEM SAFETY CRITERIA

Hazard Severity

MIL-STD-882 establishes system safety criteria guidelines to assist in the determination of hazard severity. The hazard severity categories listed in Table 2.1 provide a *qualitative* indication of the relative severity of the possible consequences of the hazardous condition(s). Although this system was initially established for use with DOD system safety efforts, it is generally applicable to a wide variety of industries that currently employ the system safety discipline. The utilization of the hazard severity categorization technique is extremely useful in attempting to qualify the relative importance of system safety engineering as it applies to a given system condition or failure. For example, the criticality of addressing a Category I, catastrophic hazard, is much more important than a negligible, Category IV hazard.

Hazard Probability

The hazard probability levels listed in Table 2.2 (MIL-STD-882) represent a qualitative judgment on the relative likelihood of occurrence of a mishap caused by the uncorrected or uncontrolled hazard. Here again, based upon a high probability that a situation will occur, a judgment can be made as to the importance of addressing one specific concern over another.

Therefore, when using the severity and probability techniques simultaneously, hazards can be examined, qualified, addressed, and resolved based upon the hazardous severity of a potential outcome and the likelihood that such an outcome will occur. For example, while an aircraft collision in midair would unarguably be classified as a Category I mishap (*catastrophic*), the hazard probability would fall into the Level D (*remote*) classification based upon statistical history of midair collision occurrence. The system safety effort in this case would require specific, but relatively minimal

TABLE 2.1 Hazard Severity Categories

Description	Category	Mishap identification
Catastrophic	I	Death or system loss
Critical	II	Severe injury, occupational illness, or system damage
Marginal	III	Minor injury, occupational illness, or system damage
Negligible	IV	Less than minor injury, occupational illness, or system damage

Source: MIL-STD-882.

TABLE 2.2 Hazard Probability Levels

Description	Level	Mishap identification
Frequent	A	Likely to occur frequently
Probable	B	Will occur several times during the life of an item
Occasional	C	Likely to occur sometime during the life of an item
Remote	D	Unlikely, but may possibly occur in the life of an item
Improbable	E	So unlikely, it can be assumed that the hazard will not occur

Source: MIL-STD-882.

controls to prevent such an occurrence. Conversely, a minor collision between two automobiles in a congested parking lot might be classified as a Category IV mishap (*negligible*) with a hazard probability of Level A (*frequent*) or Level B (*probable*). The effort here would focus on implementing low-cost, effective controls because of the high probability of occurrence. Signs indicating right-of-way, wide parking spaces, low speed limits, the placement of speed bumps, and so on, are some examples of such controls. Hence, it is fairly obvious that if evaluation of a potential for mishap reveals a Category I occurrence (*catastrophic*) with a Level A probability (*frequent*), the system safety effort would undoubtedly require elimination of the hazard through design or, at the very least, provide for implementation of redundant hazard controls prior to system or project activation.

Very simply stated: An extreme or severe hazard risk may be tolerable *if* it can be demonstrated that its occurrence is highly improbable; whereas a probable hazard may be tolerable *if* it can be demonstrated that the result of occurrence would be extremely mild. This intuitive reasoning leads to the assumption that the probability of a hazard risk is inversely proportional to its severity.

System safety hazard analysis, as discussed in this text, is concerned primarily with the identification and control of hazard probability and severity of a given project, system, or program. In fact, analysis and evaluation of system hazards are the very basis of the system safety effort. Proper analysis performed during the total life of a project will provide the essential foundation upon which the entire safety program should be based. Chapter 4 will demonstrate that adequate identification and control of hazards in the early stages of a product's life cycle will dictate the nature and extent of such standard industrial tasks as personnel training, preventative maintenance, procedure development, purchasing requirements, engineering approaches, and product design criteria. It must also be emphasized that, in general terms, system safety must examine *all* levels of operating hazard associated with a given system including the results of any potential failures. However, since some risk of hazard or accident exists even when certain systems or tasks operate as intended and designed (pressure systems, foundry operations, oil refinement, etc.), the total hazard level must be evaluated, and not just that associated with system or subsystem failures. Having established this concept of total hazard evaluation, the reader should now understand that the system safety effort would not be complete if all elements of operational integrity are not evaluated.

The Hazard Risk Matrix

Table 2.3 shows the Hazard Risk Matrix which incorporates the elements of the Hazard Severity table and the Hazard Probability table to provide an effective tool for approximating acceptable and unacceptable levels or degrees of risk. By establishing an alphanumeric weighting system for risk occurrence in each severity category and level of probability, one can further classify and assess risk by degree of acceptance. Obviously, from a systems standpoint, use of such a matrix facilitates the risk assessment process. It should be noted that Table 2.3 provides only an example of a Hazard Risk Matrix for illustrative purposes and for demonstrating the approach

TABLE 2.3   Example of a Hazard Risk Assessment Matrix—Values Can Be Assigned Based Upon Organization Preferences

Risk Assessment Matrix	⇓ HAZARD SEVERITY CATEGORIES ⇓			
FREQUENCY OF OCCURRENCE (PROBABILITY) ⇓	I Catastrophic	II Critical	III Marginal	IV Negligible
(A) Frequent	1A	2A	3A	4A
(B) Probable	1B	2B	3B	4B
(C) Occasional	1C	2C	3C	4C
(D) Remote	1D	2D	3D	4D
(E) Improbable	1E	2E	3E	4E
HAZARD RISK INDEX				
Risk Classification	Risk Criteria			
1A, 1B, 1C, 2A, 2B, 3A	Unacceptable, changes must be made			
1D, 2C, 2D, 3B, 3C	Undesirable, make changes if possible			
1E, 2E, 3D, 4A, 4B	Acceptable with management review			
3E, 4C, 4D, 4E	Acceptable without review			

to risk assessment as used in this text. The Matrix can be adjusted and modified to meet the objectives of any given enterprise or operational parameters. Table 2.3 provides four categories of severity and five categories of probability and, therefore, it is often referred to as a "4 × 5 Risk Matrix." However, some organizations will sometimes add a fifth severity value such as "insignificant" or "slight" or "no loss." In such cases, it would be referred to as a "5 × 5 Risk Matrix." The point is, the exact parameters and/or categories assigned are not written in stone and as long as the categories used are well-defined and understood by the users, the Matrix is an extremely useful tool in the evaluation of risk. Table 2.3 also shows an example of how a shaded code can be used to further highlight the categories of risk; in this example, a dark gray, medium gray, light gray, and white shade scheme has been applied. Again, organizations should customize their Matrix to meet the objectives of their specific risk assessment approach.

System Safety Precedence

The order of precedence for satisfying system safety requirements and resolving identified hazards is not unlike that which applies to general industrial safety considerations. There are five basic steps, as follows (MIL-STD-882):

1. Design for minimum risk
2. Incorporate safety devices

3. Provide warning devices
4. Develop procedures and training
5. Acceptance of residual/remaining risk

1. **Design for Minimum Risk**: The system safety order of precedence dictates that, from the first stages of product or system design, it should be designed for the elimination of hazards, if possible. Unfortunately, in the real world, this is not always practical or feasible. If an identified hazard cannot be eliminated, then the risk associated with it should be reduced to an acceptable level of hazard probability through design selection.

   To clearly understand the relative importance of this element in the system safety order of precedence, consider the following example.

   An entrepreneur wishes to establish a small manufacturing facility that will be involved in the production of school desks. Part of the finishing process will require several coats of lacquer to be applied to each desk surface. An enamel-based paint will also be used on the under-structure of each desk. The facility will have only one small open-faced paint booth. Ventilation will be provided and the operator will be supplied with respiratory protection in the form of disposable respirators. However, during the design phase, a system safety evaluation of the painting process required the identification of hazards associated with all aspects of this task, including materials/chemicals planned to be used. The analysis of the operation reveals that the designated lacquer to be used contains an isocyanate derivative, which is extremely hazardous and will require an expensive supplied-air respiratory protection system. Because a system safety analysis of this operation was performed *during the system design phase* of this project, the management of this enterprise can choose to design the hazard out of the system by selecting a less hazardous but equally acceptable paint product. If the owner wished to eliminate the potential exposure all together, an automated paint application system could be evaluated with regard to risk-reduction benefits versus cost. The obvious point here is to demonstrate that utilization of the system safety order of precedence allows management more choices in the management of risk associated with their operations.

2. **Incorporate Safety Devices**: If identified hazards cannot be effectively eliminated or their associated risk adequately reduced to acceptable levels through system design, that risk should be reduced through the use of engineering controls and safety devices. These may include fixed, automatic, or other protective safety design and hazard limitation/control features or devices. Also, when applicable, provisions should be made for periodic functional checks and maintenance of any safety devices.

   In the above example, the management of this manufacturing plant has determined that many other comparable paints/lacquers available on the market also contain isocyanates or other equally hazardous commodities. The installation of automated technologies will be too cost prohibitive to operate a competitive enterprise. Therefore, the system safety order of precedence dictates that suitable safety devices should be installed to control the hazard risk posed by the toxic lacquer. This would mean that the management team must decide whether to install a permanent supplied-air system or provide a portable, self-contained breathing apparatus to be worn by the operator only when using the hazardous paints. Physical barriers can be installed to preclude entry into the area by other plant personnel during the painting operation. Again, proper consideration of the system safety analysis process provides management a choice of hazard-control/risk-reduction techniques.

3. **Provide Warning Devices**: When neither design nor safety devices/ engineering controls can effectively eliminate identified hazards or adequately reduce the associated risk, devices should be employed to detect the condition and produce an adequate warning signal to alert personnel of the hazard. Warning signals and their application should be designed to minimize the probability of personnel reacting incorrectly to the signals and should be standardized within similar types of systems to avoid further confusion.

   Continuing with the above example, it has been determined that the design of the paint booth could not be changed adequately enough to eliminate or control the risk potential imposed by the hazardous chemical to an acceptable level. Also, requiring a paint booth operator to wear a new type of breathing apparatus carries some additional risk of noncompliance by the operator, especially when the system is new and unfamiliar. There are other company personnel in the facility not assigned to the paint operation but who are required to work in the same general vicinity within the facility. They too could possibly be exposed to some levels of toxic isocyanate vapors. In this instance, the order of precedence dictates that warning devices be installed as a further or added precaution for hazard/risk control. Such devices include, but are not limited to, warning signs posted in the operating area to remind of the hazards and/or the required use of personal protective equipment, a warning light or beacon which will be activated whenever the painting operation is in progress to preclude the possibility of other company personnel entering the area, or a public address announcement made throughout the facility to let people know when the hazardous operation starts and stops.

4. **Develop Procedures and Training**: Where it is impractical to eliminate hazards through design selection or adequately reduce the associated risk with safety warning devices, administrative controls, such as procedures and training, should be used to advise personnel how to safely operate the hazardous system. For example, procedures may include the use of personal protective equipment as a means of protecting personnel from a hazardous condition. Also, certain hazardous tasks and activities may be deemed critical and might require personnel to be certified as proficient. It should be noted that, without special consideration, no warning, caution, or other form of written advisory should be used as the only method of risk reduction for Category I or Category II hazards.

   Once again, our example, to ensure the paint booth operator is aware of the changes made to the system (new form of respiratory protection, additional warning signs, concern for other employees during paint spraying applications, familiarity with the exact hazardous nature of the toxic paint, etc.), specific operating instructions and training procedures must be developed. By ensuring adherence to an approved, written operating procedure through proper training, the potential for operator error can be further reduced to acceptable levels. The possibility of exposure to other personnel not associated with this task is also reduced through awareness training and procedural controls.

Through proper and detailed consideration of the system safety order of precedence, the potential risk of the paint operation will be reduced to its lowest perceivable level and the risk acceptance, the next and last step, will be much easier to justify.

5. **Risk Acceptance**: Realistically, even when operating in compliance with the minimum standards established by applicable safety and health regulations, there may still be some level of residual risk which must inevitably be accepted. How much risk is accepted or not accepted is a management decision. The outcome of that decision will be affected by numerous inputs and considerations, not the least of which is cost.

The process for reducing risk created by a hazard is illustrated in Figure 2.2. It reflects an obvious interaction of both engineering and management considerations to bring about an optimal resolution of risk. The final acceptance or rejection of residual risk becomes a decision of the managing activity.

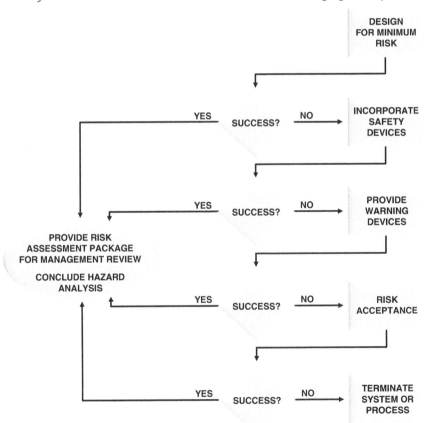

Figure 2.2   *Hazard reduction order of precedence process flow.*

COST AND RISK ACCEPTANCE

Of primary concern to management is, and will always be, the issue of cost. As an example, Figure 2.3 is a graphical illustration model of an expected loss index based upon cost of system loss versus the probability of that loss (Olson, undated). An arbitrary limit is set on acceptable mishap cost with an index of five (in actuality, any index could be used, it would just alter the slope of the line accordingly). It should be emphasized that the example in Figure 2.3 is only concerned with system loss. Personnel loss is not an issue in this example. If it were, the importance of system loss as it relates to cost would, of course, be overruled by the importance of the preservation of human life. In this hypothetical illustration, a system designed such that the probability that a mishap can occur with one chance in a thousand (10^{-3}) would be acceptable if the loss were $5000 or less. Similarly, if a loss of five million dollars were projected, a probability of occurrence of once chance in one million (10^{-6}) would be acceptable risk. Using this concept as a baseline, quantitative and qualitative design limits can be adequately defined. However, as risk/cost trade-offs are being considered through the design phase of a project, it sometimes becomes evident that certain safety parameters force higher program risk. From the management perspective, a relaxation of one or more design parameters may appear, on the surface, to be advantageous when considering the broader issue of cost and performance optimization. A facility or operation's manager will frequently make such decisions against the recommendation of the System Safety Staff. The System Safety Manager must recognize the right of the upper echelon to exercise management prerogatives when costs are involved. However, the prudent facility manager will also realize that a decision to alter design

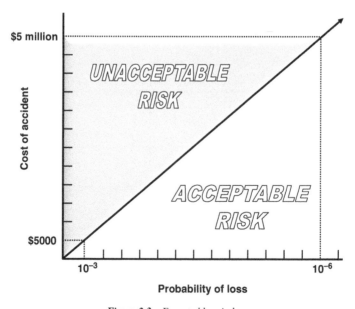

Figure 2.3   *Expected loss index.*

parameters rather than fix a safety concern must be documented properly. When a management decision is made to accept a level of risk, the decision should be coordinated with all affected organizational elements and then documented so that in future years, everyone will know and understand the elements of the decision and why it was made. When personnel loss must be considered, this documentation becomes especially critical. It will be extremely difficult to justify or even explain that the cause of some future loss of human life or limb was due to a previous decision to accept the risk based purely upon monetary cost savings. Such actions are the foundation of successful personal injury and wrongful death litigation.

Another aspect of cost as it relates to risk acceptance is the subsequent costs associated with either controlling or eliminating the risk. Some hazards are considered unacceptable, even if they pose relatively low risk, because they are somewhat easier to control/fix. For example, even though the risk of being struck by lightning, which has been calculated in the area of one in 14 million, can be considered relatively low, people seldom remain outdoors during a lighting storm. The risk here, although negligible, is worth eliminating based upon the cost of ignoring the possibility altogether (death or serious physical injury). The cost to control or eliminate this risk potential may also be minor in most cases (i.e., one could simply remain indoors). However, if a major construction operation is to remain on a tight schedule, costs of reducing personnel exposure to lighting strikes are viewed from a different perspective. In fact, many construction site managers often find themselves weighing the low risk potential of a possible lightning strike against the serious impact potential of a slipped schedule and/or cost overruns.

Conversely, there are other hazards that are considered acceptable, even though they may pose high-risk potential, but they are relatively difficult to fix. An example here would be space shuttle launch operations. From a purely system operation point of view, the level of risk associated with launching/landing a space shuttle is several orders of magnitude greater than operating an airline flight, and the risks involved in an airline flight are several orders of magnitude greater than the risk of piloting a small single-engine aircraft. Hence, cost is not only a major consideration of risk acceptance, but it also plays an important role in the evaluation process associated with risk identification and control (Olson, undated).

Because of the relative ease in obtaining data, some analysts may be tempted to assess risk in terms of the average cost of past accidents. However, this method often results in a gross underestimation of system risk. Accident patterns are random events and the average cost is usually larger than the most frequently occurring cost. This is because the very large or catastrophic accident may (and frequently does) constitute a significant portion of the total risk, even though no such accident may have occurred in recent history (DOE SSDC-11 1982).

Quantitative Risk Assessment

In any discussion of risk management and risk assessment, the question of quantified acceptability parameters must be considered. Richard E. Olson (undated) provides the following discussion pertaining to quantitative risk assessment.

In any high-risk system, there is a strong temptation to rely totally on statistical probabilities because numbers seem an easy way to measure the safety and likelihood of failure/loss. However, the limitations and basic principles of such an approach, as well as past engineering experience, should be well understood before attempting any such measurement. Quantitative acceptability parameters must be well defined, predictable, demonstrable, and, most important, useful. They must be useful in the sense that they can be easily converted into design criteria. Many factors considered fundamental to system safety are not, in actuality, quantifiable. Design deficiencies are not easily examined from a statistical standpoint. Additionally, it is entirely possible for system safety analysts and managers to become so enamored with the statistics that simpler and more meaningful methods to address a concern might be overlooked. Caution here cannot be overemphasized. Arbitrarily assigning a quantitative measure for a system creates a strong potential for the model to mask a very serious risk. Having established this understanding, it should be reiterated at this point that Figure 2.3 is only an example of how such models can be used to determine loss expectations based upon cost of system loss versus the probability of that loss. It is general in nature and care should be taken when attempting to apply this exact model to more specific systems.

In the design of many high-risk systems such as nuclear power facilities or weapon systems, there is often a strong tendency to rely solely upon statistical analysis for hazard evaluations. Management finds such an approach somewhat easier to accept since it provides a convenient (if not entirely realistic) medium to express safety in terms to which they can relate. However, the unwary can be easily trapped in their failure to establish reasonable limits on the acceptability of a probability of risk occurrence.

For example, Richard Olson: One such "high-risk" program considered a calculated probability of risk occurrence of 10^{-42} to be an unacceptable level. To illustrate the impracticality of this decision, this level of risk will be considered in terms that all can relate to—*money*. If it can be assumed that a single dollar bill is three thousandths of an inch thick, the probability of selecting that same bill from a stack of dollar bills three inches high (or 1000 dollars), is 1×10^{-3} (or 1 chance in 1000). One million dollar bills create a stack 250 feet tall. The chance of selecting that same single dollar bill from this stack is now 1×10^{-6} (or one chance in a million). When the chance goes to one in a billion, or 1×10^{-9}, the stack of dollar bills is now over 47 miles high. One chance in a trillion (1×10^{-12}) creates a stack 47,000 miles high! If probability in this example is spoken in terms of 1×10^{-42}, the stack of dollar bills probably would not fit within the confines of the galaxy. The probability of an undesired event expressing 1×10^{-42} approaches one occurrence in many times the life of the universe. The point is that realistic, reachable safety goals must be established so that management can make intelligent, rational decisions based upon understandable data. In this particular instance, the safety analysis dwelled upon the probability of the impossible and allowed a single human error, with the probability of occurrence in the range of 1×10^{-3}, to cause a near disaster, mainly because it was not a quantifiable element. It is doubtful whether the decision makers were fully aware of the mishap risks they were accepting. Instead, they were overwhelmed by a large, impressive-looking number (Olson, undated).

Principles of Risk Management

According to Olson, there are 12 generally accepted principles of risk management. A related discussion of these principles can also be found in the Department of Energy's *Risk Management Guide* (SSDC-11 1982).

a. All human activity involving a technical device or process entails some element of risk.

b. Every discovered hazard does not require panic; there are ways of controlling them.

c. Problems should be kept in the proper perspective.

d. Risk should be weighed and judgments made according to knowledge, experience, and company need.

e. Other company disciplines or organizational elements should be encouraged to adopt the same philosophy.

f. System operations represent some degree of risk; good analyses will identify the need to reduce the odds of occurrence.

g. System safety analysis and risk assessment do not eliminate reliance on sound engineering judgment.

h. It is more important to establish clear objectives and parameters for risk assessment than to find a standardized "cookbook" approach to problem solving.

i. There is no "best solution" to a safety problem or concern. There are a variety of directions in which to proceed, each of which may produce some degree of risk reduction.

j. Advising a designer on methods of achieving a specified safety goal is much more effective than indicating a suggested design will not work.

k. Total safety is a condition which seldom can be achieved in a totally practical manner.

l. There are no "safety problems" in system planning or design. There are only engineering or management problems which, if left unresolved, can cause mishaps.

MANAGEMENT COMMITMENT

System safety success cannot be achieved without firm management commitment, regardless of the nature of the business or industry. There must be a mutual confidence between company managing directors and system safety managers. Upper-level managers must have confidence that safety decisions are made with professional competence. System safety managers must know that their actions will receive full management support. Personnel must have well-defined assignments for the system safety tasks, and the authority and management flexibility to perform their assignments. Additionally, there must exist a control and coordination which will establish, in advance, what is considered an acceptable level or risk; who has resolution authority; what organizational elements should be involved; what output is required/expected; and what will be done with that output (Olson, undated).

Perhaps of primary importance in the management equation is that decision makers must be fully aware of the risk(s) they are taking in making their decisions. The system safety effort is designed to facilitate this requirement. Decision makers must then plan and manage their risk. For effective risk management, Olson suggests that responsible managers should:

a. Demand that competent, responsible, qualified engineers are assigned within the organization, as well as in any contractor organizations, to manage the System Safety Program;

b. Ensure system safety managers are properly placed within the organizational structure so that they have the authority and organization flexibility to perform effectively;

c. Ensure that acceptable and unacceptable risks are defined specifically and documented, as a company operating policy, so that decision makers are made aware of the risks being assumed when the system operates;

d. Require an assessment of mishap risk be presented as part of any program evaluation/review and as a part of all decision-making milestones.

Without the above assurances in place, as a minimum commitment from organizational management, the system safety effort will not succeed. It can be said that the very reason system safety is utilized is to facilitate the decision-making process regarding risk or potential risk of failure. Therefore, management must not only provide the necessary resources and company-wide commitment needed to accomplish the system safety objectives, it must also stand ready to accept the results of the system safety process and ensure that appropriate, responsible decisions are made based upon all available information.

3

System Safety Program Requirements

THE SAFETY CHARTER

In any organization concerned about the safety of personnel, systems, products, or services, there is one fundamental principle that must be clearly established and understood for the safety effort to succeed: *The Safety Charter*. This necessary Charter has been presented in a variety of ways over the years by numerous experts and professional consultants. However, the fundamental philosophy behind the Safety Charter has remained constant and is presented and discussed here. In a typical Line and Staff organization, the task of *safety* is most always a Staff function. This means that, while professional safety personnel are responsible for providing recommendations and advise to assist Line managers in their efforts to comply with applicable rules and regulations, it is still the Line managers and supervisors that have the authority and responsibility to implement the recommendations of Staff organizations such as safety. Having established this principle concept, the task of *safety* should be approached with the following basic understanding of the Safety Charter:

> It is essential that the Safety Function be implemented as a Line responsibility. The Safety Organizational element within the company is a Staff function which provides advice and assistance to the Line in their efforts to comply with all established safety requirements in daily operations of the organization. Safety, as a task, must clearly be the function of the Line, or Safety will not succeed.

This Safety Charter allows for safety to be a productive and functioning element of an organization's daily operations. It demonstrates that effective safety management,

Basic Guide to System Safety, Third Edition. Jeffrey W. Vincoli.
© 2014 John Wiley & Sons, Inc. Published 2014 by John Wiley & Sons, Inc.

including the system safety effort, requires not only full commitment from all levels of management, but also full management participation. Only after establishing the Safety Charter as a basic ground rule for operations can an effective system safety program be implemented.

The Safety Charter is based upon a fundamental concept which stipulates that Line management (especially first-line supervisors, but includes management from the top of the organization on down) are absolutely responsible for all operations which occur within their assigned area(s). There are very few Line managers/supervisors that would argue this position or have it any other way. It is therefore logical to add that this responsibility must include the safety of those operations. This is an extremely important concept which must be understood and accepted through all levels of the organization. Hence, the system safety effort requires managing directors, project engineers, design engineers, and so on, to ensure safety objectives are fulfilled as a given system, product, or project is conceived, designed, developed, and implemented. System safety cannot succeed if it is approached without such assurances.

SELLING SAFETY TO MANAGEMENT

It should also be noted here that in practice, the Safety Charter, as fundamental as it may be, is often a difficult concept for some organizational elements to accept. More often than not, the occupational safety function of an organization must also engage in exotic marketing strategies within their own company to literally *sell* the safety program to upper management. Unfortunately, this may also be the case when system safety programs are proposed for implementation. With system safety, however, there is a slight advantage. If approached properly, implementation of a system safety program can be shown as a cost savings strategy in the long term. The very concept behind system safety is to identify hazards within a system or process *prior to* a mishap, incident, or system failure and provide recommended solutions, corrections, or controls to preclude any such problems. Since incidents, mishaps, accidents, and/or system failure all equate to lost revenues and subsequent reduction in profits; there should be relatively little difficulty in gaining management acceptance of a properly proposed system safety effort.

In contrast, occupational safety and health programs can be more difficult to implement, especially when upper management has not established such programs as a required operating objective. For example, Wellness Programs, Safety Incentive Programs, Accident Prevention Strategies, Off-the-Job Safety Promotions are all basic to the occupational safety and health effort. While these programs have proven to be quite effective in gaining employee acceptance and boosting morale, it is often difficult to prove to company comptrollers, as well as a skeptical management, that the absence of such programs would have made any real difference in the overall safety performance of the operation. After all, how does one demonstrate how many accidents or lost-time injuries the company would have experienced *without* any of the often costly safety program elements discussed above? This question, of course, cannot be answered with any degree of certainty and is therefore only posed in an attempt to demonstrate the value of a system safety program. The point is system safety can be sold to upper management if properly proposed. In fact, it is suggested here that gaining management acceptance for system safety might possibly be less difficult than obtaining approvals for some of the most basic elements of a well-rounded occupational safety and heath program.

THE SYSTEM SAFETY EFFORT

As previously discussed, an important aspect of a successful system safety program is to ensure maximum reduction of the risk associated with a given system, product, or process produced within a given enterprise. However, an equally important element of consideration is to require exactly the same assurances from subcontracting organizations that provide any systems, products, or processes to a contracting company. As discussed in the previous chapter, system safety has its roots in the military and other government agencies responsible for its development over the past six decades. Therefore, to further understand this "federal connection" and how system safety actually becomes a required element of government contract acquisitions, the following discussion will focus on the system safety process as it relates to government contracts. Once understanding of this process has been firmly established, the reader should be able to adapt usable elements of this contracting process when attempting to implement a system safety requirement for subcontracting organizations as well as their own company.

Historically, the requirement for a system safety program has usually been the result of some sort of government acquisition. As presented in Figure 3.1, a government agency that desires a new system, product, program, or service usually establishes system safety requirements and standards at the onset of the acquisition process (i.e., the pre-bid phase). Requirements for a system safety program will be outlined in a Request for Proposal (RFP). In the RFP, the government establishes specific performance criteria which are commonly referred to as a Statement of Work (SOW). The potential contractors then "bid" their effort in accordance with the requirements established in the specified SOWs. Most always, the contract will require the bidders to implement a System Safety Program and provide a System Safety Program Plan (SSPP) which defines the methods by which the contractor intends to perform the system safety effort.

Routinely, the government will require the SSPP to contain, at the very least, the items specified in MIL-STD-882. The SSPP will typically include explanations of the contractor's intended system safety program effort. The SSPP will usually provide detailed information about the system safety personnel and their qualifications, which must meet the minimum requirements of the RFP specifications. Information pertaining to intended Standard Operating Procedures (SOPs) and other types of operating instructions are also described. The SSPP should provide data regarding required products and services which will be developed during the contract period.

The contract will also require specific products to be delivered to the customer at specified time periods or intervals. These items are usually found on the Contract

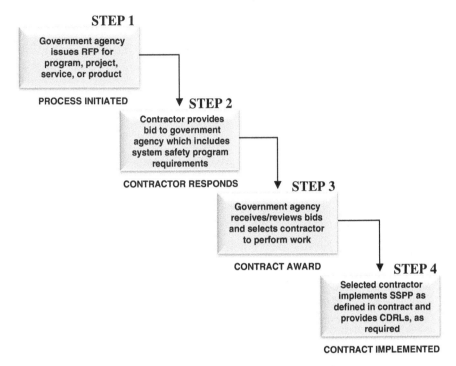

Figure 3.1   Typical system safety program process flow.

Deliverable Requirements List and are referred to as "CDRL" (pronounced *SEE-DRULL*) Items. Quite typically, the customer will require certain System Safety CDRL Items throughout the life of the contract. In fact, the SSPP itself is usually one of the first CDRL requirements. In some instances, depending upon the nature of the proposed contract, the SSPP might be submitted along with the contractor's response to the RFP. This will give the customer an opportunity to review the contractor's intended system safety program from the onset. In addition to the SSPP, system safety CDRLs may include, but are not necessarily limited to, the following items:

• Closed-loop hazard tracking system/plan
• Accident risk assessment
• Mishap/accident/incident reporting
• Facility inspection reports
• System safety analyses

Closed-Loop Hazard Tracking System

The government agency/customer will require that the contractor implement a system for identifying, tracking, and closing hazards associated with contractual operations. These previously unforeseen or unknown hazardous conditions may develop as the result of the operation of a specific facility, equipment, hardware, or a combination of these. As indicated in Chapter 2 (Figure 2.1), all of the elements in the working environment, including people, must be considered when attempting to identify hazards to a task, job, or process. Once a hazardous condition has been identified, it should be documented on some sort of Hazard Report. Figure 3.2 is a sample Hazard Report form which can be used to document as well as track the corrective action/closure status of such hazards. Completing the Hazard Report initiates the tracking process. The Closed-Loop Hazard Tracking System requires the contractor to provide documented evidence to the customer indicating that each of the identified hazards has been effectively closed or controlled to an acceptable level of risk so as not to be a threat to normal operations. The customer is able to provide a response indicating approval or disapproval of the closure or control actions which "closes the loop" and ensures complete accountability for the safety of the system. Figure 3.3 shows how an identified hazard is incorporated into the Closed-Loop system, tracked, controlled or closed, and reported back to the customer for approval.

Accident Risk Assessment

In addition to the hazardous conditions which develop during daily operations and are incorporated in the Closed-Loop Hazard Tracking System, the customer may require the contractor to perform a formal periodic risk assessment of all facilities in which operations will occur. This is done usually annually, but it can be more frequent if the customer so desires or if operational activities dictate. The risk assessment will also take into consideration the hazards associated with the permanent equipment and hardware assigned for use in the facility. The Accident Risk Assessment then becomes a detailed safety analysis of a facility, its systems, and functions. It provides the customer a single source of reference for information regarding a specific area of operations. Depending upon the depth of the assessment, it can also be a valuable tool when changes or modifications to a facility are required. A good risk assessment will provide enough detailed information about the current operating configuration of a facility or system and will, therefore, facilitate customer review and approval of any proposed modifications. Of course, after any significant modification or change to an existing system, the Accident Risk Assessment should be updated accordingly and submitted again.

In short, the Accident Risk Assessment provides a comprehensive, detailed evaluation of the overall accident risk associated with the operation and maintenance of a specific facility, its systems, equipment, and hardware. It incorporates the results of integrated hazard analyses, recommended design changes, hazard reports, and procedural or administrative tools which will eliminate or reduce the risk of these hazards, operational flow charts, safety-critical procedure lists, and other such information pertinent to the overall assessment of accident risk.

Mishap/Accident/Incident Reporting

The necessity to report mishaps, accidents, and/or incidents to the contracting agency should, at face value, be obvious. In fact, the occurrence of such unfortunate activities

HAZARD REPORT FORM

SYSTEM OR ACTIVITY		NUMBER	REV
CLOSURE STATUS ☐ OPEN ☐ CLOSED		**DATE**	

HAZARD DESCRIPTION

ACTION TAKEN

VERIFICATION

APPROVAL SIGNATURES

——————————————— ——————————————— ———————————————
SYSTEM SAFETY MANAGER   PROJECT MANAGER   CUSTOMER REPRESENTATIVE

Figure 3.2   *Sample hazard report form.*

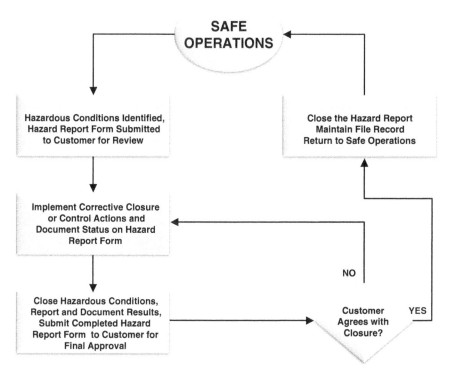

Figure 3.3   *Typical closed-loop hazard tracking system flow.*

may provide new or modified interpretations of previous risk assessments. However, not so obvious is the method by which a contractor determines which occurrences are considered "reportable" and which are not. For this reason, and because the contracting agency usually wishes to avoid inundation of paperwork for *every* incident (major and minor), the contract will typically specify conditions or limits which, if met or exceeded, will require the submittal of a formal report. For example, the US Air Force will follow reporting criteria as established in Air Force Regulation 127-4, "Investigating and Reporting US Air Force Mishaps." Among other things, this document basically requires the contractor to report, "… without delay, any accident/incident to Government property in excess of $1,000.00, hospitalization of one or more employees and any fatality …" This information is provided here as an example of the military criteria used in mishap reporting. In the private sector, organizations are free to establish their own internal mishap reporting criteria. With such preestablished guidelines identified, the contractor is better able to determine which accidents, incidents, mishaps, and so on, require reporting to the contracting agency. Also, the customer may require submittal of detailed lessons-learned and corrective action intentions along with the Accident Report. Since one of the primary objectives of the system safety effort is to eliminate or reduce accident risk potential through design and/or control actions, it is absolutely essential for the system safety function to play an integral part in the accident reporting and lessons learned process. If the subject accident, incident, mishap, and so on, was the result of previously unknown or unforeseen hazardous conditions, then a system safety reevaluation is necessary to preclude the possibility of future, similar events and to ensure optimum control of system operations.

Facility Inspection Reports

Periodic, scheduled facility safety inspections are essential in any operational area, especially where hazardous tasks are performed on a regular basis. Compliance with safety inspection requirements should not be difficult to accomplish since similar requirements should already exist in an established occupational safety program. The facility inspection encompasses all facets of daily operations and considers the human–machine interface a primary candidate area for potential mishaps. Frequent facility inspections are an excellent method of maintaining current awareness of facility conditions and how those conditions affect, or might affect, the safe operation of that facility. A system should be in place to ensure implementation of corrective actions and to track repetitive items. Results of inspections should be properly documented and accountability for discrepant items appropriately determined and assigned for the inspection process to be effective. If properly performed, the facility safety inspection is an excellent tool in the overall success of the system safety function.

System Safety Analyses

The contract may require a wide range or types of system safety analyses to be performed for a variety of reasons during the life of the contract. For example, any time new equipment or hardware is introduced into the work environment, a series of system safety analyses should be performed. Likewise, when existing equipment is modified to the extent that critical functions of the equipment may be affected, a series of analyses should be conducted prior to the first operational use of the modified equipment. In addition, prudent system safety protocol will dictate that certain analyses be conducted under certain circumstances. For example, an accident investigation may utilize fault tree analysis, or the system safety technique known as MORT (Management Oversight and Risk Tree) to determine the exact cause(s) that lead to an accident/incident/mishap.

LIFE CYCLE PHASES AND THE SYSTEM SAFETY PROCESS

Any proposed or existing project or product has what is called a *life cycle*. Within the project life cycle are sub-elements known as *phases*. Each distinct phase will, in turn, indicate certain tasks that are typically performed in the life of that project. These tasks are required to establish, implement, and maintain a successful system safety process. Generally, the tasks fall within three broad categories:

1. **Planning Tasks**: Tasks needed to initiate a new program such as the development of policy, operating requirements, expected results of the effort, schedules, and so on, are part of the planning task. These tasks are typically the responsibility of upper management and are usually performed by the system safety staff or, if established, a system safety planning group. Review and subsequent approval by company management is also required as part of the planning task.
2. **Primary System Safety Tasks**: The tasks of identifying, initiating, and controlling hazards are part of the primary system safety task. These efforts are performed by a variety of contributors including management, safety, engineering (the design element) and, in most cases, the end user of the product.
3. **Support Tasks**: Such tasks needed to maintain the program include, but are not limited to, training, documentation, and database generation and are normally assigned to the safety staff.

Figure 3.4 shows the typical life cycle of a generic project with the various phases within the life cycle identified, as well as the system safety program elements, as follows:

Concept Phase

During the concept phase of the life cycle, overall project goals and objectives are identified and developed. Project descriptions, design requirements, and expected end results are also formulated. In effect, the concept phase establishes a project *road map* to provide direction and purpose to the proposed project. A loosely translated Chinese proverb provides a very good justification for the elements of the concept phase: "… if you don't know where you are going, any road will take you there!" A properly executed concept phase will not only identify the project destination, but it will also keep it on the desired track through completion.

Design Phase

The general plans developed during the concept phase are made more specific during the design phase. Plans, drawings, design specifications and parameters, and so on, are clearly developed during the design process. Once project design(s) are reviewed and approved by all participating organizational elements, the project can then move into the next phase with confidence that the end result will be as expected with very few, if any, real surprises along the way.

Production Phase

All the conceptual planning and approved design criteria are finally transformed into the desired product during the production phase. Since actual production is taking place along with greater expenditures of time and money necessary to produce a physical product, the production phase is justifiably considered by many system

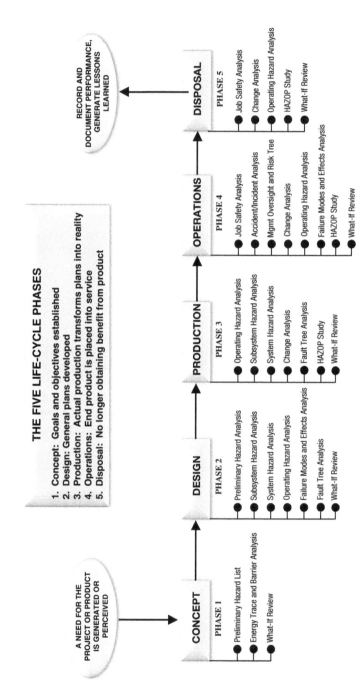

THE FIVE LIFE-CYCLE PHASES

1. **Concept:** Goals and objectives established
2. **Design:** General plans developed
3. **Production:** Actual production transforms plans into reality
4. **Operations:** End product is placed into service
5. **Disposal:** No longer obtaining benefit from product

A NEED FOR THE PROJECT OR PRODUCT IS GENERATED OR PERCEIVED

RECORD AND DOCUMENT PERFORMANCE, GENERATE LESSONS LEARNED

CONCEPT

PHASE 1

- Preliminary Hazard List
- Energy Trace and Barrier Analysis
- What-If Review

DESIGN

PHASE 2

- Preliminary Hazard Analysis
- Subsystem Hazard Analysis
- System Hazard Analysis
- Operating Hazard Analysis
- Failure Modes and Effects Analysis
- Fault Tree Analysis
- What-If Review

PRODUCTION

PHASE 3

- Operating Hazard Analysis
- Subsystem Hazard Analysis
- System Hazard Analysis
- Change Analysis
- Fault Tree Analysis
- HAZOP Study
- What-If Review

OPERATIONS

PHASE 4

- Job Safety Analysis
- Accident/Incident Analysis
- Mgmt Oversight and Risk Tree
- Change Analysis
- Operating Hazard Analysis
- Failure Modes and Effects Analysis
- HAZOP Study
- What-If Review

DISPOSAL

PHASE 5

- Job Safety Analysis
- Change Analysis
- Operating Hazard Analysis
- HAZOP Study
- What-If Review

Figure 3.4   Project life cycle phases and the system safety process.

38

safety professionals as one of the more critical phases of a project life cycle. For this reason, a great deal of effort is required by all responsible/participating organizational elements to ensure project success.

Operations Phase

The end product is put into operation during this phase. Whether the end product is a facility, a piece of operating equipment, a tool, or a service, the entire effort of the system safety process will be realized during and throughout the operations phase. If the end user is an internal organization, the system safety professional has an opportunity to closely observe product operation and to make subsequent evaluation of the risks associated with that operation. However, if the end user is an external agency, there is seldom an opportunity to conduct operational evaluations of risk. Either way, *the proof is in the pudding* and the system safety efforts prior to the operational phase should therefore be as complete as possible.

Disposal Phase

The beneficial use of the end product has reached a point of diminishing return and is destroyed, discarded, or the operation is discontinued. Depending upon the nature of the product, this phase may require disposal planning, phase-out deconfiguration, actual tear-down or disassembly, and so on. Such actions are especially necessary when dealing with hazardous products or systems such as hazardous waste treatment equipment/facilities, underground storage tank removals, and nuclear power plant modifications.

Part II of this text will detail a number of the various common system safety analytical methods and techniques that are practiced in the system safety discipline. Each of these methods or techniques are usually conducted at specific points during the project/product life cycle, as indicated in Figure 3.4. At this point, it is important to understand that a specific system or program may require the use of any or all of the system safety analyses techniques available to today's system safety professional. Each method has its own distinct purpose and function, and, as tools, each can be quite useful.

It should be clearly understood that, although system safety programs are usually the required result of the government acquisition process, the discipline of system safety and its concepts, tools, techniques, and so on, can and should be utilized in nongovernment programs and/or contracts. A variety of system safety reference documents, procedures, operational guides, and handbooks have been developed as a result of six decades of work by government system safety engineers and professionals. There is every reason to borrow and transfer this knowledge into the private sector, and industry is encouraged to do so. The benefits from an implemented comprehensive system safety program far outweigh the potential costs associated with unsafe operations/tasks, equipment, facilities, and hardware. As long as people are required to perform in a work environment filled with any or all of these elements, the importance of a system safety program is even more evident.

4

The Industrial Safety Connection

THE OCCUPATIONAL SAFETY AND HEALTH ACT

Safety and health professionals who work in industrial settings concentrate primarily on ensuring compliance with, at the very least, the minimum safety and health standards promulgated under the Occupational Safety and Health Act of 1970, as enforced by the Occupational Safety and Health Administration (OSHA). These rules form the basis of almost all occupational safety and health programs currently in place throughout both the public and private sectors.

Safety and health regulations for general industry can be found in the US Code of Federal Regulations (CFR) at 29 CFR 1910 and, for the construction industry, at 29 CFR 1926. These codes address literally thousands of situations, working conditions, hazard control requirements, and worker safety and health protection standards, all intended to make the work environment safer. Although numerous areas are addressed and a minimum level of worker protection is ensured by implementation of these many standards, it is important to note that the major driving factor for worker safety is contained not in the CFR but in Section 5 of the Occupational Safety and Health Act itself. This factor, better known as the *General Duty Clause*, Section 5(a)(1), simply states that:

> Each employer shall furnish to each of his employees employment and a place of employment which are free from recognized hazards that are causing or are likely to cause death or serious physical harm to his employees, and shall comply with occupational safety and health standards promulgated under this Act.

Basic Guide to System Safety, Third Edition. Jeffrey W. Vincoli.
© 2014 John Wiley & Sons, Inc. Published 2014 by John Wiley & Sons, Inc.

It is clear that the intention of the Occupational Safety and Health Act is to ensure a safe and healthy workplace. In fact, the General Duty Clause has been the precursor for many of the regulations that have followed since 1970 and continues to ensure worker safety in the absence of a specific standard for a specific occupational situation. Less clear, but equally important to worker safety, is that the methods and techniques associated with the system safety effort are an excellent means of assuring that the intent of the General Duty Clause is met at the earliest possible time in the project or program development process.

Historically, the occupational safety and health profession has relied primarily on achieving "safety" through compliance with established standards and mandated operating criteria. However, as noted earlier, most standards and regulations reflect *minimum* requirements for safe operations and do not place any additional responsibilities upon employers to be *safer* than what is required by these standards. It is well known within the occupational safety arena that, in order to ensure continued safety in a given situation, it is often necessary to exceed the minimum requirements established by law.

Another primary goal of the industrial safety effort is the reduction or elimination of occupational accident potential. Here again, implementation of the system safety process can provide a defined means of accomplishing this objective. Even though a majority of industrial accidents/incidents have historically been attributed to the unsafe acts or the unsafe conduct of workers, the importance of unsafe physical conditions and equipment can and should not be minimized in any discussion concerning accident risk potential. Injuries and/or property damage caused by mechanical hazards generally have a high potential severity, since they often result in a permanent partial or permanent total disability (loss of motion or use of a body member, amputations, loss of sight, damage to hearing, etc.) and/or extensive damage or loss to essential equipment or facilities. Furthermore, many of the so-called *unsafe acts* that cause such injuries or damage may not result in an accident if the potential for risk is properly assessed well in advance and safer physical conditions are implemented to control the hazards associated with the level of ascertained risk. The concept of system safety analysis provides an excellent opportunity for the industrial safety practitioner to achieve the desired goal of an accident-free work environment.

The very concept of system safety, as established in previous chapters of this text, is to

a. Systematically evaluate and analyze a given project, process, product, facility, service, and so on, to identify the risk of hazard associated with that system and

b. Recommend/implement risk elimination or control techniques so that management can make intelligent and informed decisions to reduce the risk of hazard to the lowest possible levels of acceptance.

It has been suggested here that this basic system safety concept is in direct correlation with the stated intentions of the Occupational Safety and Health Act, as established by the General Duty Clause. In many cases, then, system safety cannot only be the means by which the much desired compliance with OSHA requirements can be achieved, but actual improvements to an industrial safety process can also be established.

It is strongly suggested that the system safety function and the industrial safety process be closely integrated to ensure a complete and sound hazard control, worker protection, and risk management program. The overall importance of frequent and positive interaction between those responsible for occupational safety and system safety cannot be overstated. The employee and the working environment do play a significant role in management's determination of acceptable levels of risk. Therefore, accurate evaluation of applicable occupational and environmental standards and regulations, as well as an analysis of specific worker tasks are absolutely essential in determining appropriate levels of protection. Since the goals and objectives of both the industrial safety and system safety disciplines tend to serve each other's best interests, it would not be prudent management practice to ignore the integration of the system safety program with the industrial safety effort.

THE HUMAN FACTORS ELEMENT

Perhaps one of the strongest argument for the integration of system safety with industrial safety programs is the *human factors element*. The social concerns for workplace safety that began in the early part of the last century, which eventually led to and then became the essence of the Occupational Safety and Health Act of 1970, are still the driving force behind the OSHA regulatory process. In fact, the industrial safety movement has evolved from a primary concern for the preservation of human life and limb. Therefore, to fully comprehend the relationship between system safety and industrial safety, one must also understand how system safety can be successfully utilized in the analysis of the human factors element of task performance. In the design of equipment, for example, human factors or *ergonomics* is a subject of great consideration. One reason for such emphasis has been the desire to design systems with extremely high levels of reliability. The desire to achieve "total reliability" in system design is not only dependent upon the equipment, but also on the way that equipment is operated by the human element. Therefore, the design of the system must be in such a way so as to ensure that the human operator can interface with the equipment in an effective manner with minimum opportunity of error. If this basic concept of human–system interface is not properly considered in the design phase, then all the safety incentives and motivational programs money can buy will not encourage an individual to operate poorly designed equipment at a designated level of effectiveness. Also, if personnel are trained to operate inadequately designed equipment under normal operating conditions, then they will typically revert to very ineffective operations under emergency or other stress-induced conditions.

Another significant aspect of the human factors equation which should not be overlooked centers around the issue of product liability, especially in the commercial world of sales and service. The concept of *strict liability* has been the basis for numerous legal judgments during the past few decades. This philosophy implies that the liability for the use and, more importantly, the misuse of products can be extended to the designer, manufacturer, and the seller of those products. This *higher degree* of responsibility for the use and misuse of a product requires the product or equipment designer to have a greater knowledge of the human factors element.

In short, it is essential that the design of a product or system consider the people–equipment interface during the very early stages of the design process if the final product is to have a high degree of reliability.

ACCIDENT PREVENTION THROUGH SYSTEM DESIGN

In the study of ergonomics, the concept of *typical human behavior* based upon documented evaluations of human performance, has provided strong evidence that certain aspects of expected behavior can potentially lead to unsafe acts. These data suggest that the design engineer can effectively reduce or eliminate serious risk concerns through proper design considerations based upon expected, or "norms" in human behavior. It is important to understand that there is no real evidence to suggest that a "normal" or "average" human being exists since so many variables are involved in the evaluation of human performance in any given task. However, for the purpose of simplification and example, the reader is provided some examples of common behavioral traits that can be expected from a large percentage of the population under most circumstances. Examples of this human behavior–design consideration process are provided in Table 4.1. One can see that simple analysis of the human/system interface in the design phase will effectively identify serious hazard risk potential resulting from this expected human behavior. Once such risk potential is identified, controls can be designed into the system. These studies in human behavior strongly suggest that optimum design safety must allow for the equipment or machine to perform in the most effective manner possible while design consideration is also given to allow the human operator to perform in the safest manner possible. Any trade-offs between this optimum design situation must consider the consequences of failure if the system does not function as intended. Effective system design, then, depends on the designer's evaluation of those areas where humans can do the best job and, conversely, those areas where safer task performance is achieved if the machine accomplishes the action. These concepts are the basis of industrial safety accident prevention programs. It is suggested that system safety can augment these accident prevention efforts, if used properly. Chapter 8 will further discuss the ergonomic element as it relates to the operational analysis of the people–task interface.

Assuring safety of a system, process, or operation and the prevention of failures (or accidents) through design has always been a cornerstone objective in the practice of system safety. However, the concept of preventing accidents and incidents through system design has only become an issue of focus and concern in mainstream engineering relatively recently. In the United States, for example, the National Institute for Occupational Safety and Health (NIOSH) led a national initiative referred to as Prevention through Design (PtD) to promote this concept and highlight its importance in all business decisions.

TABLE 4.1   Samples of Behavior Patterns That Must Be Considered During System Design

Human behavior considerations in system design	
Behavior description	Design consideration
People do not **USUALLY** consider the effects of surface friction on their ability to grasp and hold and article.	Design surface texture to provide friction characteristics commensurate with functional requirements of task or device.
MOST people cannot estimate distances, clearances, or velocities very well (they tend to overestimate short and underestimate large distances).	Design so that users do not need to make estimates of critical distances, clearances, or speeds. Provide indicators of these measurements when necessary.
MOST people do not watch where they place their hands and feet, especially in familiar surroundings.	If hand/foot placement is critical to the process, design so that careless, inadvertent placement of hands or feet will not result in injury. Provide guards, restraints, warning labels, etc.
People **OFTEN** utilize the first thing available as an aid in getting where they want to go or to manipulate or "fix" something.	Either design the product so that the "first thing handy" simply cannot be functionally useful, or so that it will serve a specific, intended function.
People **SELDOM** anticipate the possibility of contact with sharp corners or edges.	Except to meet functional requirements, eliminate sharp edges on surfaces or units where inadvertent human contact is even remotely possible.
People **RARELY** consider the possibility of fire or explosion from overheated objects.	Unless it is an unavoidable functional requirement, eliminate configurations that will permit such potential (even with product misuse).
MANY people do not take the time to read labels or instructions, or to observe safety precautions.	Make labels brief, bold, simple, and clear. Repeat or place same labels on various parts of a product. Make use of color coding, fail-safe innovations, and other attention-demanding devices.
MOST people perform in a perfunctory manner, utilizing previous habit patterns. Under stress or in an emergency, they almost always revert to these habit patterns.	Do not alter or change an established design (if it is satisfactory) just for the sake of change. Base all "design innovations" on changes in functional or operational requirements.

Source: DOE.

According to NIOSH, the concept of PtD can be defined as: *Addressing occupational safety and health needs in the design process to prevent or minimize the work-related hazards and risks associated with the construction, manufacture, use, maintenance, and disposal of facilities, materials, and equipment.* A growing number of business leaders have recognized PtD as a cost-effective means to enhance occupational safety and health. Many US companies openly support PtD concepts and have developed management practices to implement them. Other countries are actively promoting PtD concepts as well. In 1994, the United Kingdom began requiring construction companies, project owners, and architects to address safety and health during the design phase of projects, and companies there have responded with positive changes in management practices to comply with the regulations. Australia developed the Australian National OHS (Occupational Health and Safety) Strategy 2002–2012

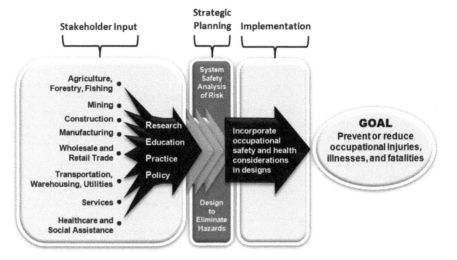

Figure 4.1   *The U.S. NIOSH Prevention through Design (National Initiative) concept, modified to show system safety integration into the process.*

that set "eliminating hazards at the design stage" as one of the five national priorities. As a result, the Australian Safety and Compensation Council (ASCC) developed the Safe Design National Strategy and Action Plans for Australia encompassing a wide range of design areas including buildings and structures, work environments, materials, and plant (machinery and equipment).

The approach used to develop and implement the PtD National Initiative is framed by industry sector within four functional areas: *Research, Education, Practice, and Policy.* As shown in Figure 4.1, this process encourages stakeholder input through a sector-based approach. The ultimate goal of the PtD initiative is to prevent or reduce occupational injuries, illnesses, and fatalities through the inclusion of prevention considerations into all designs that impact workers. During this process, intermediate goals are identified to provide a path toward achieving the ultimate goal. In the United States, NIOSH serves as a catalyst to establish this initiative but, in the end, the partners and stakeholders must actively participate in addressing these goals to make PtD "business as usual" in the twenty-first century.

A major goal for the NIOSH PtD Plan as a National Initiative was the development and publication of a new National Standard to provide guidance on including PtD concepts within an occupational safety and health management system that can be applied in any occupational setting. In 2011, the American Society of Safety Engineers (ASSE) announced the release of the American National Standards Institute (ANSI)/ASSE Z590.3 Standard, "Prevention through Design: Guidelines for Addressing Occupational Risks in Design and Redesign Processes." The Z590.3 standard focuses specifically on the avoidance, elimination, reduction, and control of occupational safety and health hazards and risks in the design and redesign process. Through the application of the concepts presented in the standard, decisions about occupational hazards and risks can be incorporated into the process of design and redesign of work areas, tools, equipment, machinery, substances, and work processes. Design and redesign also include construction, manufacture, use, maintenance, and disposal or reuse of equipment used on the job. One of the key elements of this standard is that it provides guidance for "life-cycle" assessments and a design model that balances environmental and occupational safety and health goals over the life span of a facility, process, or product. The Z590.3 focuses on the four key stages of occupational risk management: the preoperational, operational, postincident, and postoperational stages are all addressed. The standard also provides tools for determining and achieving acceptable levels of risk for hazards that cannot be eliminated during design.

Z590.3 complements, but does not replace, performance objectives existing in other specific standards and procedures. The goals of applying PtD concepts in an occupational setting are to

- Achieve acceptable risk level;
- Prevent or reduce occupationally related injuries, illnesses, and fatalities;
- Reduce the cost of retrofitting necessary to mitigate hazards and risks that were not sufficiently addressed in the design or redesign processes.

According to NIOSH, implementation of the standard can save lives and prevent injury. For example, as skylights become synonymous with green construction and energy conservation, an increase in skylight installation can be expected. If skylights are designed and installed with proper guarding, deaths and injuries to workers who inadvertently fall through skylights during construction and maintenance activities could be prevented. Another example involves bailing machines used to break down cardboard for recycling in various industries. If the bailers were designed and installed with proper guarding, workers would not be able to enter the machines for trouble shooting thus preventing deaths and injuries.

The American Industrial Hygiene Association (AIHA) is also actively engaged in including PtD concepts into the revisions to (ANSI)/AIHA Z10 "Occupational Health and Safety Management Systems" standard. Including PtD into Z10 can be considered a natural progression since the "systems" approach to occupational health and safety enables management and worker collaboration on corrective actions, enables the anticipation of occupational hazards so that risks to workers from those hazards can be assessed and controlled during design, and enables the inclusion of occupational safety and health into an organization's planning process.

Further research and investigation into the contents and requirements of both the Z359.3 standard and the Z10 standard is highly recommended as a next step in the study of the PtD approach to accident prevention and its relationship to the system safety process.

THE PROCESS OF TASK ANALYSIS

Industrial or occupational safety and health professionals have been involved for some time now in the analysis of tasks which must be performed in the workplace and the human interface which must occur in order to accomplish such tasks. In actuality, the modern methods of task analysis and job analysis were initially developed as a result of the somewhat historic time and motion studies conducted in the early part of the twentieth century. These techniques were later enhanced further during the US Department of Labor's efforts in occupational analysis conducted in 1930. These methods represent an analytical process which moves from the detail of individual physical job movements to the generalized description of the occupation itself. Later, during World War II, the need for an even more complete analysis of human performance was first perceived. This need lead to the development of the formal *task analysis* process which is nothing more than a thorough examination of the individual elements and sub-elements that comprise a given task (DOE SSDC-31 1985). It is a systematic review of a collection of actions or human behaviors necessary and sufficient to complete that task. To fully consider the human element and its effect on the operation of a given system, it is necessary to analyze the specific operational task requirements of the particular system in question. The process of task analysis provides the means by which this transition from the more general human factors studies to the more specific human factors considerations which have been customized to a particular system for the sole purpose of designing an effective human–task or human–machine interface, as the case may be.

Task analysis differs from Job Safety Analysis in that the latter is a more simplified, global, higher function description of a job and its corresponding tasks assigned to one person and related to general safety.

Task analysis outputs may serve a variety of system inputs. For example, task analysis may be used to obtain detailed information on a given work position, thereby providing data for selection, training, staffing level, procedure development and retrofit, communications, equipment review, feedback, supervisory control, and risk screening.

THE JOB SAFETY ANALYSIS AND SYSTEM SAFETY

The *Job Safety Analysis* or JSA (also referred to as the *Job Hazard Analysis* or *JHA*), which is a more simplified form of task analysis, has been a long-standing tool for task and function analysis. JSA has been available and utilized in general industry for many years by the industrial safety community. However, many practitioners do not understand or are simply unfamiliar with the connection between the JSA and the system safety tasks of hazard identification and analysis. It has even been suggested by some in the profession that the JSA itself is a type of oversimplified system safety analysis and, if performed earlier in the job development phase, could be used as the basis of a Preliminary Hazard Analysis for a specific task or set of tasks. However, because JSA is often (and improperly) used only to analyze a function *after* it has been implemented, much of the data is not factored into the system safety process. The primary purpose of the JSA is to uncover inherent or potential hazards which may be encountered in the work environment. This basic definition is not unlike that which has been previously discussed regarding the various system safety analyses.

The primary difference between the two is subtle but important and is found in the end-use purpose of the JSA. Once the job or task is completed, the JSA is usually used as an effective tool for training and orientating the new employee into the work environment. The JSA presents a verbal picture of a specific job, broken down into a step-by-step description of the tasks required to perform that job. Therefore, since the JSA attempts to identify hazards inherent to an *existing job*, its primary purpose then is not pretask hazard elimination or risk reduction (as is the case in system safety analysis), but hazard control through awareness and training. Also of importance when attempting to differentiate the JSA from the system safety process is to point out that the JSA is typically performed by the person or group of persons most knowledgeable of the specific job or task, usually the supervisor, lead technician, and/or department head. This, of course is not the case with a system safety analysis. While it is true that a supervisor or department head may have valuable input in the system safety effort, seldom are they the only contributors to the system safety evaluation process.

To further illustrate the important difference between the JSA and the system safety process, consider the primary elements of the basic Job Safety Analysis, which typically include the following five steps (as a minimum):

1. Select a job;
2. Break the job down into steps;
3. Identify the hazards to determine the necessary controls of the hazards;
4. Apply the controls to the hazards;
5. Evaluate the controls for effectiveness.

The JSA, then, is a specialized approach of task analysis that takes an existing job and analyzes its tasks to specifically identify hazards encountered in the work environment. At the very least, the JSA does have a place within the system safety process as a tool to evaluate the hazards/risks of an existing task or function during the *operation phase* of the project life cycle. Here we see another connection between the principle elements of the industrial safety process and one of the basic objectives of the system safety effort: The JSA tries to eliminate or control the risk of hazard exposure in a given task during the life of the project.

As a minimum, the expected benefits of a properly performed JSA can be summarized as follows with respect to the principle concept of system safety:

- Provides individual training in the safe, efficient operation of equipment/hardware;
- Establishes positive safety contacts with employee work force;
- Provides new personnel with "on-the-job" safety awareness training;
- Prepares for planned safety observations during the performance of the task and;
- Provides prejob safety instructions for irregular or nonroutine tasks.

Figure 4.2 shows a typical Job Safety Analysis form which can be utilized by any organization wishing to capitalize on this basic method of hazard identification and analysis. It should be noted again that it is most ideal if the task supervisor completes each JSA for those operations under his/her direction. This makes sense since it is the supervisor who is most likely to have detailed knowledge of the steps associated with each task (including the hazards). Also, by including those personnel in the JSA process who will actually perform the work, even more valuable insight will be gained and a more complete understanding of job hazard control will inevitably be realized.

GUIDELINES FOR PREPARING A JOB SAFETY ANALYSIS

The JSA and the analysis of job/task risk is (or should be) a critical element in the assurance of worker safety and health. However, its potential for success can be severely hindered when the JSA is not utilized or performed properly. As stated earlier in this chapter, it is usually the task supervisor and the work team that will complete the JSA. At the very least, this means that consistency in JSA approach and completion can be as varied as the tasks being analyzed. Ideally, the safety professional should also participate in JSA development to facilitate the process and ensure the proper and complete analysis of the given task. But, in reality, most industrial safety and health practitioners may not always be involved or even present when a JSA exercise takes place.

In an effort to provide some basic guidelines for conducting a JSA, the following information is provided. It might be suggested that the JSA is the closest link the industrial safety world has to the principles and concepts of system safety analysis. Establishing these guidelines at this juncture will hopefully demonstrate the parallel objectives of industrial safety assurance and system safety analysis.

Figure 4.1 shows a JSA form that can be used in performing the safety analysis. Figure 4.3 provides an example of a completed JSA form for reference purposes only. The JSA form is divided into three columns. Instructions for completing each column are provided on the first page, over the respective column. It is extremely important to follow these instructions closely to ensure a properly completed and usable JSA. This cannot be overemphasized. Specifically:

Column 1 (Sequence of Job Steps): The job shall be broken down into specific steps describing, in sequence, what is to be done. The descriptions must be clear, simple, and concise (usually no more than one or two brief sentences). Important things to remember when filling information in Column 1:

- Describe *only actual job steps* that contain a hazard(s), create a hazard(s), or expose personnel to a hazard(s), or hazardous condition(s). Avoid placing information in Column 1 (or anywhere else on the JSA) that does not specifically address an actual job step or task. Non "step-specific" or general safety information can be found and documented elsewhere (the contractor's Health & Safety Plan, Company Directives, Tailgate Briefings, etc.) and does not belong in the JSA.

Job Safety Analysis

Job Description:		JSA No.:
Job Location:		Contractor Name:
Page ____ of ____	Prepared By: (Originator's Name)	Date:
Signature – Team Members		

SEQUENCE OF BASIC JOB STEPS	POTENTIAL ACCIDENTS OR HAZARDS	RECOMMENDED SAFE JOB PROCEDURE
Break the job down into basic steps that tell what is done first, what is done next, and so on.	Ask yourself for each step, what accidents could occur to the people during the job step.	For each potential accident, ask yourself what exactly should that person do or not do to avoid the accident.
Record the job steps in their normal job order of occurrence.	Ask: Can they be struck by or contacted by anything? Can they be caught in, on, or between anything? Can their fall? Can their strain or over exert themselves? Can they be exposed to gas, fumes, radiation, etc.?	Describe specific precautions in concrete detail. Give each recommended precaution the same number as was given each job step to which it applies. List the procedure number for those job steps included in the procedure.
Describe what is done, not the details of how it is done. Usually three or four words are sufficient to describe each job step.		Avoid generalities like "be alert," "be careful," and "take caution."
Make the job steps neither too fine nor too broad. They should sound natural. Sometimes the job step may be a major safety precaution; e.g., "Check for gas before entry."		Use simple do and don't statements. If necessary, explain how, as well as what to do.
Number each step.		Also, question the basic job method. Is there an entirely different way to do the job that is better and safe? If a repair or service job, can anything be dome to increase the life of the job?

Figure 4.2 Typical Job Safety Analysis (JSA) form.

51

Job Safety Analysis
Page 1 of 3

Job Description: Prepare Plant Site for Energy Project (SAMPLE)		JSA No.: SAMPLE
Job Location: Anywhere, USA (SAMPLE)		Contractor Name: ABC Contractor, Inc (SAMPLE)
Prepared By: (Originator) SAMPLE		Date: SAMPLE
Signature – Team Members SAMPLE	SAMPLE	SAMPLE

SEQUENCE OF BASIC JOB STEPS	POTENTIAL ACCIDENTS OR HAZARDS	RECOMMENDED SAFE JOB PROCEDURE
Break the job down into basic steps that tell what is done first, what is done next, and so on.	Ask yourself for each step, what accidents could occur to the people during the job step.	For each potential accident, ask yourself what exactly should that person do or not do to avoid the accident.
Record the job steps in their normal job order of occurrence.	Ask: Can they be struck by or contacted by anything? Can they be caught in, on, or between anything? Can they fall? Can their strain or over exert themselves? Can they be exposed to gas, fumes, radiation, etc.?	Describe specific precautions in concrete detail. Give each recommended precaution the same number as was given each job step to which it applies. List the procedure number for those job steps included in the procedure.
Describe what is done, not the details of how it is done. Usually three or four words are sufficient to describe each job step.		Avoid generalities like "be alert," "be careful," and "take caution."
Make the job steps neither too fine nor too broad. They should sound natural. Sometimes the job step may be a major safety precaution; e.g., "Check for gas before entry."		Use simple do and don't statements. If necessary, explain how, as well as what to do.

Also, question the basic job method. Is there an entirely different way to do the job that is better and safe? If a repair or service job, can anything be dome to increase the life of the job? |
| Number each step. | | |
| 1. Installation and maintenance of corrosion control measures (including, swales, silt fencing, and seeding). | 1a. Slip/Trip/Fall due to various uneven surfaces and ground areas (dips, pits, ditches) and debris in the area. | 1a1. Personnel shall observe and note obstacles that are in their path of travel, taking care to avoid areas of uneven footing and/or debris that could cause them to trip or fall.

1a2. Housekeeping at site shall occur at end of each shift to remove unnecessary debris and avoid accumulation of clutter and/or supplies |
| | 1b. Struck by heavy equipment (dozer, back hoe) during swale construction. | 1b1. All personnel entering area shall wear high visibility vests at all times.

1b2. Equipment operators shall maintain a maximum speed of 5 mph while in work areas. |
| | 1c. Eye contact/injury with foreign objects/debris due to dusty/windy conditions. | 1c1. All personnel must wear approved PPE, including safety glasses with side shields. |
| | 1d. Contact with snakes, insects, animals, rodents (including rodent waste/feces), and other wastes. | 1d1. All personnel shall be briefed on the presence of biological hazards and shall avoid contact with animals and insects during work activities. |

(Continued on next page)

Figure 4.3   Example of a partially completed JSA form.

Job Safety Analysis

Page 2 of 2

SEQUENCE OF BASIC JOB STEPS		POTENTIAL ACCIDENTS OR HAZARDS		RECOMMENDED SAFE JOB PROCEDURE
2.	Demolition of existing structures and fencing	2a.	Cuts/scraps/abrasions while working with or handling demolished fencing materials.	2a1. All personnel working with demolished fencing materials shall wear work gloves of substantial material (leather)and long -sleeved shirts.
		2b.	For access to elevated areas: If **ladder** is used, possible fall due to unstable ladder footing and/or overreaching by user during cleaning operations causing user to loose balance or ladder to fail	2b1. Whenever an employee is using a ladder, a second employee will serve as spotter, supporting the base of the ladder and ensuring proper ladder footing at all times.
		2c.	For access to elevated areas: If **elevating boom lift** is used, possible fall due to travel over uneven floor surfaces or debris	2c1. All travel will occur with boom or scissors lift in lowest possible position.
				2c2. Only personnel that have been trained and certified to operate elevating and rotating man-lifts will be permitted to use the equipment.
				2c3. All removable debris shall be cleared of the equipment path of travel prior to movement
		2d.	Sprains/strains while manual lifting of heavy loads.	2d1. All personnel shall practice safe lifting techniques, seeking assistance for loads that are too heavy to lift alone.
3.	Clearing, grubbing, stripping, and disposal of vegetation, including tree removal	3a.	Cuts/scraps/abrasions while working with sharp tools used during clearing/grubbing operations.	3a1. All personnel involved in clearing/grubbing operations shall wear work gloves of substantial material (leather), long-sleeved.
		3b.	Crushed toes while lifting/removing trees and shrubs.	3b1. Personnel shall be required to practice safe lifting techniques and wear appropriate work shoes (steel toed safety shoes preferred).
4.	Installation of access roads, parking, and lay down areas	4a.	Struck by heavy equipment (blades, pavers, etc.) during pavement operations	4a1. All personnel shall be required to wear high-visibility vests while in operational areas.
				4a2. Heavy equipment operations will not occur unless a spotter or signalman is available to direct tracffi and keep area clear.
				4a3. All heavy equipment shall be equipped with functional back-up alarms which will be tested prior to each work shift.

Figure 4.3 *(Continued)*

- Number each step in sequential order (i.e., 1, 2, 3, 4, etc.). This will allow easy reference to a particular step, especially in a multipage JSA. It will also facilitate association of hazard descriptions (in Column 2) and mitigation measures (in Column 3) to the steps described in Column 1.

- IMPORTANT: While there may be 50 steps to a given job, only two of those steps may actually be hazardous. Therefore, the *only steps* that are to be described in Column 1 will be the two hazardous steps. In other words, do NOT place a job step in Column 1 unless there is at least one hazard associated with the performance of that step. Adherence to this requirement is sometimes difficult, but it is essential to ensuring a properly completed JSA. Remember, a JSA is NOT an "operating instruction."

- The JSA only addresses hazards to *people* (i.e., the person performing the step and/or those around him/her). This cannot be overemphasized. The JSA does NOT consider hazards to property or the environment, unless those hazards also threaten personnel. If the latter is the case, then only the threat to personnel shall be addressed on the JSA.

- If a job step cannot be described in one or two brief sentences, then reevaluate the step that is being considered. It is possible that it can be further broken down to be more descriptive of the task. Keep the descriptions simple and clear.

Column 2 (Potential Accidents or Hazards): For each specific step described in Column 1 (Sequence of Job Steps), provide a description of the hazard(s) and/or hazardous condition(s) associated with the performance of that step in Column 2. Important things to remember when completing the information in Column 2:

- In many cases, there will be more than one specific hazard associated with a given job step. For this reason, each hazard will be numbered alphanumerically to associate the hazards with the proper job step. For example, if job step number 2 contains three specific hazards, then the hazard descriptions in Column 2 of the JSA form shall be numbered "2a", "2b," and "2c." In this way, it will be quite obvious that job step number 2 contains three individual hazards of concern to the worker.

- If a job step is placed in Column 1, then there MUST be at least one hazard associated with that step described in Column 2 (otherwise, the step should never be listed in Column 1). That hazard MUST be clearly described (see next bullet). Do NOT refer readers to other sections of the JSA for a hazard description. Do NOT refer readers to other documents (e.g., another JSA, a regulation) for a description of a hazard that is supposed to be specific to a given job step. Just describe the hazard, plain and simple.

- When describing the hazard(s), be specific. What is it about a particular step that presents a hazard to personnel (i.e., what is the hazard)? It may be helpful to ask questions, focusing on the following key American National Standards Institute (ANSI) hazard categories:

– Can a person be *struck by* something (e.g., moving equipment, vehicles, flying debris)?

– Can a person *strike against* something (e.g., sharp edges, fixed equipment)?

– Can a person *fall to the same level* (e.g., slip/trip)?

– Can a person *fall to a different level* (e.g., working at heights)?

– Can a person be *caught in, on, or between* something (e.g., pinching or crushing hazards)?

– Can a person *overexert* themselves (e.g., sprain/strain/lifting)?

– Can a person *come into contact with* something hazardous (e.g., electricity, heat, cold, radiation, caustics, dusts/fumes/vapors/mists/gases/smoke, noise/vibration, toxic or noxious substances, biohazards)

– Can a person be *placed into hazardous locations* (e.g., confined spaces, poorly illuminated areas, work in tight spaces)?

• Remember to only describe the *hazard(s)* in Column 2. Avoid describing control measures here. It is often difficult to do, but save the control measures (the mitigation) for Column 3 of the JSA form. Also, avoid further describing the job step in Column 2. All descriptions of the job step belong in Column 1 of the JSA form.

Column 3 (Recommended Safe Job Procedures): For each specific hazard described in Column 2 (Potential Accidents or Hazards), provide specific mitigation measures that will effectively ensure either the elimination of the hazard (preferred approach) or maximum exposure control. Important things to remember when filling information in Column 3:

• In many cases, there may be more than one recommended control measure for a particular hazard. For this reason, each control measure will be numbered alphanumerically to associate the control with the proper hazard (and job step). For example, if a hazard associated with job step number 2 has been labeled "2a" and there are two possible or required control measures for hazard 2a, then the control measures in Column 3 of the JSA form shall be numbered "2a1" and "2a2" respectively. In this way, it will be quite obvious that hazard number 2a for job step number 2 contains two individual control measures that must be implemented (2a1 and 2a2).

• Be specific to the hazard when describing control requirements. *Always* list the requirements for *each* hazard associated with *each* job step. Avoid referring readers back to previous steps when identifying control measures, even if such measures have been previously described for some other step. This can become confusing and, in some cases, lead to misinterpretation (i.e., the reader refers to the wrong control requirements). It is preferable to be repetitious rather than confusing when describing hazard control measures.

• Never use vague or nondescript terminology such as "be safe" or "be careful" or "take caution." These are *not* control measures and do not provide the reader with any usable information.

- Use simple "do" and "do not" statements whenever possible. If necessary, describe how and what to do, as well as what not to do (as the case may be).
- Avoid referencing Company Directives, Regulations, and/or Training Courses as the *sole means* of hazard control. This is practically useless to the reader. Rather, describe the required control measure(s) for the hazard(s). It is highly improbable that the reader will search out a Directive or Regulation and attempt to locate/identify the proper portion of the Directive that applies to the specific hazard. Remember, the JSA is intended to provide a worker with the information he/she needs to perform the work safely. Help them in every way possible. Make the JSA easy to understand and to use.
- If a hazard is described in Column 2, then there MUST be at least one control/mitigation measure associated with that hazard described in Column 3 (otherwise, the hazard should never be listed in Column 2 and the work step never listed in Column 1). That control/mitigation measure MUST be clearly described. Do NOT refer readers to other sections of the JSA for a control measure description. Do NOT refer readers to other documents (regulations, other JSA, etc.) for a description of a hazard control measure that is supposed to be specific to a given job step hazard. Just describe the control measure, plain and simple.

Signatures and Approvals

Upon completion of the JSA, the JSA team members sign the document in the spaces provided (first page of the JSA form) to indicate their concurrence with the information presented therein.

Prior to implementing the JSA, it is recommended that the appropriate company safety representative review the completed JSA form. If circumstances warrant, a safety representative may have also been a member of the JSA team, in which case their signatures as a team member also indicates their approval of the completed JSA form.

Attachments to the JSA form:

Attachments to the JSA can include any permits or documents necessary to control the hazards specified within the JSA (e.g., Confined Space Entry Permits, Excavation and Trenching Permits).

Changes in Hazard/Scope

As described in this chapter, JSAs are normally developed for a specific activity. In the event of an unforeseen change in scope, change in hazard level, or the identification of a new hazard, the work should be suspended and the JSA be revised accordingly. The revised JSA must adequately address the hazards associated with the change(s). All appropriate signatures (i.e., representatives from those organizations or departments that signed the original JSA) should be obtained on the revised JSA prior to restart of the subject work.

SYSTEM SAFETY: AN INTEGRAL PART OF THE OVERALL ORGANIZATION

The role of the occupational or industrial safety and health organization in the system safety process has been established as an essential element since both can be interpreted as "*self-serving*" to a great extent. That is to say, the industrial safety program could be drastically improved by incorporating the process of system safety whenever possible and, conversely, a well-rounded system safety effort would not be complete without adequate consideration of the industrial safety program.

Once again, the primary objective of a system safety program within any organization is to ensure the maximum level of safety (i.e., the lowest level of acceptable risk) while operating within the boundaries of effectiveness, time, cost, and feasibility. A properly implemented system safety effort in any industrial organization will effectively apply appropriate scientific and/or engineering techniques and principles to first identify and then eliminate (if possible) or control any risk of exposure to system hazards. Therefore, it is also essential to understand that other internal departments of an industrial organization should be included in the system safety process. Without appropriate participation between organizational departments, the system safety effort might not succeed or, at the very least, might fall short of identifying all possible system risks. In a manufacturing facility, for example, possible interdepartmental system safety interfaces might include the following organizational elements/personnel:

- **Facility Engineers**: As indicated by the previous example of an intended paint booth operation at a desk manufacturing facility, the importance of system safety participation and coordination with facility engineers cannot be overemphasized. The most logical opportunity for the evaluation and elimination or control of potential hazards is when a design for a new or improved facility is in the preparation stages of development. The participation of the system safety specialist in the planning of new or modified facilities should also help to ensure a more thorough consideration of safeguards that can be built into the new development. As an example, in the layout of a new office facility, it is not unusual that only cursory consideration is given to the location and quantity of electrical receptacles for the operation of essential office equipment. Then, when the equipment is actually installed, the typical result is an unsafe amount of electrical extension cords strung precariously around desks, across floors and along aisle ways. It is often difficult (and expensive) to correct such an oversight after completion of initial construction. With proper consideration and evaluation of the risk potential associated with the work environment (i.e., people, equipment, facilities, and procedures) through the system safety process, such unfortunate and potentially unsafe circumstances could be avoided.

- **Equipment Design Engineering**: The system safety organization should provide safety criteria, design parameters, and other necessary requirements to the design engineers. System safety should also play a role in the review of system and component specifications, interface and control drawings, schematic diagrams, and so on, and provide input as required during the conceptual phase to verify incorporation of appropriate safety controls. The design engineer should also have a significant input in the performance of the preliminary hazard analysis. With proper consideration of the human factors element in the design of a new product, potential incidents involving human error might be avoided through proper design. For example, utilization of known human performance/reliability data during the design process will assist in the identification and control of human error hazard causes.

- **Reliability, Maintainability, and Quality Control**: Inclusion of these organizations in the system safety process, from concept through disposal, will aid in the identification of safety critical components for reliability analysis. A Failure Modes and Effects Analysis, as well as other common reliability models, can be used to identify critical and noncritical failure points. The Quality Assurance element can be extremely useful in the overall system safety process. Quality engineers should participate in the inspection of safety critical components, serve on certification boards, audit any corrective action requirements, and identify any safety impacts associated with implementation of such requirements.

- **Engineering Management**: Since engineering management must review and approve any engineering changes, including those driven by system safety requirements, they too play a key role in the system safety process. Also, since they are a management function, they are typically cost-sensitive and their support is, therefore, critical to the success of the system safety program.

- **Manufacturing/Facilities Operations and Maintenance**: These organizations can be very helpful in identifying hazardous materials, equipment, processes, and/or operations associated with a new or proposed product. System safety personnel should consult with occupational safety and facility engineers to determine facility safety requirements and safe operating criteria. Perhaps of greatest utility is the ability of operations and maintenance personnel to assess operating procedures and proposed work flow to identify, in advance, the potential for failure/mishap and propose methods to eliminate hazardous conflicts or situations. Maintenance personnel can assess new tool requirements, preventative maintenance provisions, any necessary new test equipment to accommodate the proposed system, and so on, and provide useful information to the system safety organization to assist in the development of the overall system safety evaluation.

- **Personnel Training Department**: Another essential participant in the system safety process is the training department. They will develop training requirements and lesson plans for the new system based upon data supplied by the design engineer, the system safety staff, and other organizations that may have been involved. Because the information provided by the trainer is usually the worker's first exposure to a new or different system, there is a unique opportunity for the trainer to ensure that one of the primary messages received by the trainee concerns the safe use (and operating hazards/restrictions) of the new product or system. The training staff can also play an active role in the development of any operating procedures which will eventually be used to establish safe and proper system utilization criteria.

* **Environmental Engineering**: If the intended, or unintended, use of the proposed system might possibly have an adverse effect on the surrounding environment, the environmental engineering and compliance function within the organization should be consulted during the system safety effort. The accurate assessment of potential environmental exposures is essential in the success of any proposed project. Possible waste management, air pollution, ground water contamination, community awareness of potential hazards, employee training for hazard awareness, and necessary emergency response activities/planning are all necessary considerations which must be assessed during the evaluation of overall system hazard and risk reduction.

It cannot be overemphasized that the principle elements of a sound industrial safety program, with its primary purpose of OSHA compliance, work hazard reduction, assurance of employee/job safety and health, and the evaluation of jobs or tasks (through the JSA or other comparable method), can, in most cases, be achieved through application of the system safety process. The connection between the two programs while not entirely obvious is quite understandable, as described above. Perhaps the most important thing to remember here is that the industrial or occupational safety and health professional can utilize the time-proven techniques of hazard reduction and system safety analysis to accomplish the desired goal of both programs:

Maximum *safety in the performance of a task or function with* **minimum** *risk of hazard exposure and* **minimum** *cost.*

5

Probability Theory and Statistical Analysis

INTRODUCTION

In the practice of modern system safety analysis, the system safety engineer attempts to provide a sufficient level of information to organizational management so that informed decisions may be made regarding hazard risk acceptance or rejection. In the safety and health arena, the provision of such choices often requires ample substantiation in order to justify decisions to accept a hazard risk. The system safety practitioner can utilize a wide variety of techniques and methods to determine risk levels and, through preestablished acceptance criteria, make recommendations to management. These analytical tools serve to *qualify* the risk in relation to some existing level and/or standard of operation. Some of the more common of these tools are discussed in detail in Part II of this text. When actual failure rate data are known or can be determined or deduced, the system safety effort can take the analysis process further and actually *quantify* the risk of hazard in terms of these known or expected failure rates.

While probability theory examines the likelihood of a specific failure event given a single opportunity for occurrence, statistics focuses on the number of times a failure event will occur given many opportunities.

Through the use of basic probability theory and statistical analysis, the system safety function can actually assign expected values to certain hazards and/or failures to determine the likelihood of their occurrence. The availability of such quantifiable

Basic Guide to System Safety, Third Edition. Jeffrey W. Vincoli.
© 2014 John Wiley & Sons, Inc. Published 2014 by John Wiley & Sons, Inc.

information further enhances the management decision-making process and justifies the existence of the system safety effort within the organization.

This chapter will present the fundamental principles of probability theory and briefly examine the use of statistical analysis in the practice of system safety. The information discussed here should provide the reader with a very basic understanding of these concepts which, by some accounts, is essential to the overall understanding of the system safety discipline. It should be noted that it is not within the scope of this *Basic Guide to System Safety* to provide all there is to know regarding probability theory and statistical analysis. However, a certain level of understanding is essential and will therefore be discussed here.

PROBABILITY

The theory of probability, as it applies to system safety, is based upon the chance or unplanned occurrence of random failure events. In general, probability studies are primarily concerned with predicting the occurrence of uncertainties in task or operational performance. The probable or likely occurrence of these events is represented numerically as a value between zero and one. An event that has absolutely no chance of occurrence (i.e., 0% of occurrence) is assigned a probability level of zero. Conversely, those events that have a 100% chance of occurrence are assigned a level of one. Since nothing can have less than a 0% or greater than 100% chance of occurrence, probabilities cannot be represented in numbers less than zero or greater than one. There are an infinite number of possibilities for event occurrence that exist between the end points of zero and one. A probability of 0.5 would indicate that a given event has an equal chance of occurring or not occurring. Such probabilities are often used daily in weather forecasting. It is not unusual to hear that there is a 50% chance of rain, or a 30% chance of snow. These probabilities have been *deduced* from available data that enables a forecaster to *predict* the outcome of the day's weather activity. When the forecast does not materialize as advertised, it is usually the result of additional, unpredictable variables that interrupted the forecasted weather pattern.

Another example to understand probability theory: Nine clean, usable machine bolts and one defective machine bolt are all placed in an opaque parts bag. The defective bolt is the same size, weight, and color as the nine clean bolts. Only the thread pattern on the bolt is defective and this cannot be determined with the naked eye. What is the chance the machinist will choose the defective bolt in a random selection exercise? Since the position of each bolt in the bag and the selection process are completely uncontrolled factors, each of the 10 bolts has an equal chance of being chosen. The probability of selecting the defective bolt on the first draw is simply stated as a 1 in 10 chance, or 0.1 in terms of probability. If this exercise were performed numerous times, the uncertainty of this probability level becomes clear. For example, if the machinist were to continue to choose bolts 100 more times, and returning the selected bolts to the parts bag between draws, the previously assigned probability level of 0.1 indicates that, out of 100 draws, the defective bolt will be pulled 10 times (100×0.1). However, in actual practice, it is conceivably possible (although highly improbable) that the inferior part could be pulled each of the 100 attempts or not at all. Therefore, the likelihood or probability of choosing a defective bolt occurring between 0 and 100 times can be *predicted* based only upon the known information that exists (e.g., 10 bolts, 1 defective, 9 clean, 0.1 probability of a defective pick on one draw).

Chance events such as random part failures, accidents, injuries, and so on, usually occur as a result of actions from which more than one outcome is possible. Hence, it follows that the more complex the action, the greater the probability of chance or unplanned results. Chance events can also refer to those unplanned or unpredictable events which result from a combination of or an interaction between conditions and/or activities. Since accidents are basically unplanned or unpredictable events, they can generally be analyzed using probability theory and statistics.

Mathematically, probability can be defined as the number of times an event will produce a given result divided by the total number of events in the sample. This probability value can be deduced or inferred. In practice, probabilities can be relatively exact if they have been deduced from known conditions. The toss of a coin or the choosing of a playing card from a deck of 52 cards has very definite and exact probable outcomes. Based upon known information, an *inferred* prediction can be made on how many times the coin will land on heads or tails. Statistics has often been referred to as the "mathematics of inference" since many probabilities for complex systems must be determined through statistical evaluation of known or inferred data. This logic of inference can be extended to the industrial environment when attempting to forecast failure rates. If certain performance data for a given part or component can be obtained or determined, probable failure rates can be calculated and assigned. For example, many machine manufacturers and vendors provide data on expected failure rates for critical parts such as gauges, valves, regulators, crane components, lifting slings, certain automotive parts, and so on. Such information is also used in the determination of safe operating factors. In the system safety process, this information is used to determine the likelihood of system, subsystem, or component failure and, therefore, greatly facilitates the decision-making process. Use of statistical inference will provide estimates of probability, determine the likelihood of occurrence based upon past experience, and predict with some level of accuracy how many times the event will occur in the future based upon deduced probability. For example, in the industrial safety world, one of the most common *techniques* for determining accident potential is to examine historical performance (i.e., past history). If a factory experienced an average of 20 lost-time injuries for every 1000 employees, the average injury probability is 0.02. If this probability is multiplied by the entire employee population of 7500, for example, then the expected number of injuries that will occur in the next year can be estimated at 150. Caution is warranted in applying these expected rates to individuals or to those employees that work in high-risk areas. An individual may have a higher or lower accident potential depending on job duties, training, experience, and so on. Likewise, an office secretary will have presumably lower risk potential than a forklift operator or welder. Of course such variables will effect the expectation. However, from a plant-wide perspective, if the sample population from which the expected probability has been estimated is similar in general to the entire population to which this rate is being applied, then the expected probability rate is as close to a prediction as is possible based upon known data. From this point, the safety engineer can begin to focus on those higher risk areas and attempt to formulate recommendations for controlling the risk of hazard or reducing it to acceptable levels.

The simple probability theory discussed thus far has not considered the effect of other factors which usually influence the outcome of an unexpected event. For example, failure of a single component in a complex system does not necessarily mean that a failure of the total system will occur. Likewise, an automobile accident does not always mean that an injury or fatality will occur. There are usually other influencing variables that will affect the outcome. If the occupants of the vehicle were wearing seat belts, the chance of an accident resulting in injury is significantly reduced. If an airbag supplemental restraint system were installed in the vehicle, the probability of injury is lessened even further. If the vehicles involved in the accident were traveling at low rather than high rates of speed, there may be an even greater reduction in the potential risk of injury. These factors create what is referred to as *conditional probability*. Conditional probability reflects the chance of an expected outcome when considering the fact that the likelihood of that outcome has been influenced by certain conditions acting upon the initiating event.

If the probability of a failure can be calculated as something less than one (i.e., less than 100%), then it follows that the probability of success is equal to one minus the probability of failure. In a previous example, the probability of experiencing a lost-time injury was calculated at 0.02. The probability of having no lost time injuries is equal to $1-0.02$, or 0.98 (a 98% chance that no lost-time injuries will occur). This ability to provide management with projected failure AND success rates for a given operation, task, system, and so on, demonstrates the advantage of using probability theory in system or project analysis.

Chapter 11 of this text will discuss the use of Fault Tree Analysis in determining system reliability, failure potential, and even accident cause factors through examination of specific or general fault paths. Additional information on the application and use of probability values in Fault Tree Analysis will also be provided in Chapter 11.

STATISTICS

Statistics evaluate any variation in the probable number of events and attempts to define these variations. For example, if a coin is tossed, there is a 50% chance the coin will either be heads or tails (probability of 0.5). If the coin is tossed 10 times, the probable number of tails (or heads) is five. This information indicates that, if many coin tossing exercises were to occur in groups of 10, then five tails would occur most often in those groups. The frequency of four or six tails would be less, three or seven tails even less frequent, and so on. Each of these values can be plotted in terms of frequency of occurrence. In situations where the probability of *all* possible outcomes of an event is known, such as with a coin toss, it is referred to as *preassigned*

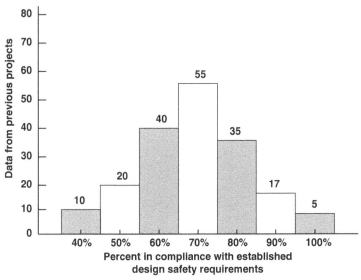

Total number of samples: 182

Probability (P) of 70% or less $= \dfrac{55 + 40 + 20 + 10}{182} = 0.687$ Probability (P) of 70% or more $= \dfrac{55 + 35 + 17 + 5}{182} = 0.615$

Probability (P) between 60% and 80% $= \dfrac{40 + 55 + 35}{182} = 0.714$

Figure 5.1   *Histogram of distribution values.*

probability. However, in the real world, not all occurrences are as predictable as coin tossing or the rolling of dice. For instance, although the probability of a desired successful event such as meeting design safety parameters may be known, many possible outcomes exist. The consideration of all possibilities when they are not known is referred to as *empirical probability* and involves statistical evaluation of possible data values. The manner in which the different values appear over the entire range of possible values is referred to as the *distribution* of values. Figure 5.1 is a histogram that demonstrates the distribution of values pertaining to the desire to meet design safety parameters based upon historical performance to date. In this example, it is easy to see how known values can be plotted, their distribution determined, and the probability of future successful performance established. Figure 5.2 takes the values from the histogram and converts the data to percent of occurrence and smoothes the points to form a distribution curve. In practice, there are many types of distribution curves, the more common of which is known as the *normal distribution* and is expressed as Figure 5.3. Note the similarities between Figure 5.3 and that of the smoothed data curve in Figure 5.2. In the industrial safety arena, these values can also be numbers of accidents, costs, injury severity rates, injury frequency rates, and so on. The most common or most frequent value that appears during evaluation or observation is known as the *mode*. If an arithmetic average is calculated by simply adding all the value points and dividing by the total number of points, the *mean* is

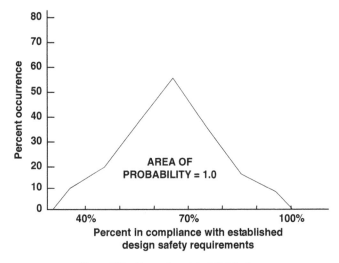

Figure 5.2   *Curve of empirical distribution.*

derived (Figure 5.3). The *median* is that value point where half of the total values under consideration lie above and the other half fall below. *Variance* is a measure of variation in the observed values and the *standard deviation* is the square of the variance.

Statistical values can also be used to determine expected periods of optimum performance in the life cycle of products, systems, hardware, or equipment. For example, if the life cycle of humans were plotted on a curve, the period of their lives that may be considered "most useful" in terms of productivity and success,

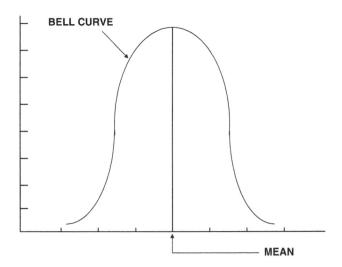

Figure 5.3   *Curve of normal distribution ("bell curve").*

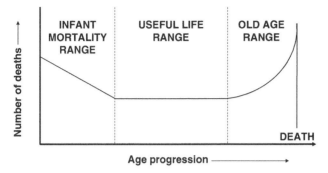

Figure 5.4   *Human life cycle curve.*

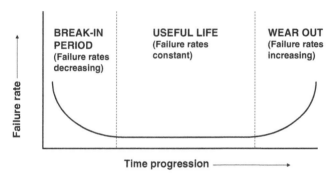

Figure 5.5   *Reliability curve ("bathtub curve").*

could be represented as shown in Figure 5.4. This plotted curve is often referred to as the *bathtub curve* because of its obvious shape. A similar curve can be used to determine the most productive period of a product's life cycle based upon the five known phases of that life cycle, as discussed in Chapter 3. The resultant curve, known as a product's *reliability curve*, would look like that which appears as Figure 5.5. During the break-in period, failures in the system may occur more frequently, but decreasingly less frequent as the curve begins to level toward the useful life period. Then, as the system reaches the end of its useful life and approaches wear-out, more frequent failure experience is likely until disposal.

SUMMARY

A detailed understanding of all statistical terms and the formulas that are associated with their use is not an essential prerequisite to the practice of *basic* system safety analysis. A familiarization with their meaning is more than adequate for this purpose. The primary difference between statistics and probability is that probability attempts to predict the occurrence of future events, whereas statistics is used to develop models based upon past performance. Using such statistical models facilitates the probability process of predicting future events. Mathematically, probability is expressed as a value between 0.0 and 1.0 with an infinite number of possibilities in between. Once a probable event has either occurred or not occurred, it becomes a statistic and the basis for future models. These models are needed because most events in the real world are considered empirically probable (i.e., many possible outcomes). The relatively few events where all probable results can be preassigned because of limited possible outcomes (tossing a coin, drawing a playing card, rolling a die, etc.) seldom require extensive statistical modeling to ensure proper calculation of those probable outcomes.

Hence, statistical evaluation of failures that occur during a product's life cycle help to develop a failure curve which, because of its shape, is referred to as a bathtub curve. When considering the usefulness of a product, the curve becomes a reliability curve for that product.

System Safety Analysis: Techniques and Methods

Part II of this *Basic Guide to System Safety* will present and briefly discuss some of the more common system safety analytical tools used in the performance of the system safety function. Through example analyses of hypothetical mechanical and/or electrical systems, the reader should become familiar with each type of system safety analysis method or technique discussed. However, it must be understood that it is not within the limited scope of this volume to provide a detailed explanation of each of these methods and/or techniques. The intention is to merely introduce the reader to the various tools associated with the system safety process. The value of each concept in the analysis of hazard risk will be dependent upon the individual requirements of a given organization or company.

It is hoped that the introductory information provided here will familiarize the reader with the primary tools of system safety analysis and provide an opportunity to experiment with these basic concepts, as they may be applied to their own safety function or responsibility.

Basic Guide to System Safety, Third Edition. Jeffrey W. Vincoli.
© 2014 John Wiley & Sons, Inc. Published 2014 by John Wiley & Sons, Inc.

6

Preliminary Hazard Analysis

INTRODUCTION

The *Preliminary Hazard Analysis (PHA)* is an analysis of the generic hazard groups present in a system, their evaluation, and recommendations for control (TAI 1989). The PHA is usually the first attempt in the system safety process to identify and categorize hazards or potential hazards associated with the operation of a proposed system, process, or procedure. In many instances, however, the PHA may be preceded with the preparation of a *Preliminary Hazard List (PHL)*. The identification of hazards on a PHL can occur through the use of a variety of methods such as but not limited to

- Checklists,
- Hazard matrices,
- The lessons learned process,
- Equipment descriptions,
- Accident/incident report data,
- Past operational history of similar tasks, and/or
- Review of other historical records.

After examining all available information, a PHL can then be prepared. Though not entirely necessary in the overall system safety program, a well-developed PHL can be expanded and further developed into the basis of a PHA. The methods used to develop

Basic Guide to System Safety, Third Edition. Jeffrey W. Vincoli.
© 2014 John Wiley & Sons, Inc. Published 2014 by John Wiley & Sons, Inc.

the PHA are similar, but more specific, than those used during the development of the more generalized PHL. Since the primary objective of the PHL is to document and provide an initial assessment of hazards identified very early in the process, the PHL should be performed and completed in the concept phase of the product or project life cycle (Stephenson 1991; Roland and Moriarty 1983). As stated, the PHL can then be used as the foundation upon which to base the PHA, and it will also assist in ascertaining the extent of the system safety effort which must follow. Also determined as part of the PHL is a preliminary *Risk Assessment Code (RAC)*. The RAC assigns or determines a risk level for a given hazard and is used to assess the scope of the system safety effort required, as well as identify the need for alternate design approaches early enough in the concept phase to be considered feasible. As stated above, the PHL can be developed from information obtained through a variety of input sources such as checklists (see Figures 6.1 and 6.2 for examples of checklists), informal meetings and conferences, and/or other previously performed analytical methods and techniques such as the *Energy Trace and Barrier Analysis (ETBA)* (Stephenson 1991). Chapter 8 will discuss the ETBA in more detail. Figure 6.3 shows a typical PHL worksheet and provides some instruction on the types of information usually recorded in the separate columns of the PHL.

The PHA Development Process

The PHA (Figure 6.4) is perhaps the most critical analysis which will be performed because it is usually the first in-depth attempt to isolate the hazards of a new or, in some cases, modified system. The PHA will also provide rationale for hazard control and indicate the need for further, more detailed analyses, such as the *Subsystem Hazard Analysis (SSHA)* and the *System Hazard Analysis (SHA)*. The PHA is usually developed using the system safety techniques known as *Failure Modes and Effects Analysis (FMEA)* (Chapter 9) and/or the ETBA. Data required to complete the PHA include, but is not necessarily limited to, any available data having to do with the following:

• Mission or scope/intent of product
• Environment in which the product will operate
• Systems concepts used to develop the product
• Equipment/hardware to be used with product
• Operational criteria for product end use

PHA development can be somewhat simplified through the use of a Preliminary Hazard Matrix (Figure 6.5) identifying a Generic Hazard Group. An example of such a group is listed in Table 6.1. For clarification, the Generic Hazard Group elements from Table 6.1 are defined as follows:

Collision: Item breaking loose and impacting other items. It hits something, or something hits it. Typically caused by structural failure, procedural error, or inadequate handling of equipment (i.e., human behavior).

GENERIC HAZARD EVALUATION CHECKLIST

SYSTEM/PROGRAM: _____

PERFORMED BY: _____   DATE: _____

HAZARDOUS ELEMENT SOURCES	YES	NO	COMMENTS
ACCELERATION			
CHEMICAL (DISSOCIATION, REPLACEMENT, SUBSTITUTION)			
ELECTRICAL ELEMENTS AND OPERATIONS			
ENVIRONMENT			
LEAKAGE			
MOISTURE			
OXIDATION			
OFF-GASSING OF MATERIAL PROPERTIES			
PRESSURE HIGH, LOW, RAPID CHANGE			
STRESS			
STRUCTURAL FAILURE			

Figure 6.1 *Sample generic hazard evaluation checklist.*

GENERIC ENERGY SOURCE EVALUATION CHECKLIST

SYSTEM/PROGRAM: _____

PERFORMED BY: _____   DATE: _____

ENERGY SOURCE (Kinetic/Potential)	YES	NO	COMMENTS
ACTUATING DEVICES			
CATAPULTED OBJECTS			
CHARGED ELECTRICAL CAPACITORS			
CHEMICAL REACTION			
CRYOGENIC MATERIAL			
ELECTRICAL GENERATORS			
ELECTROMAGNETIC, IONIZING NONIONIZING RADIATION			
EXPLOSIVE CHARGES			
ELECTROSTATICE CHARGE OR DISCHARGE			
FALLING OBJECTS			
FUEL/PROPELLANTS			
GAS GENERATORS			
HEATING DEVICES			
INITIATORS/IGNITORS			
NUCLEAR			
PLACEMENT OF SYSTEMS, COMPONENTS			
PRESSURE CONTAINERS			
PUMPS, BLOWERS, FANS			
ROTATING/MOVING MACHINERY			
SPRING – LOADED DEVICES			
STORAGE BATTERIES			
SUSPENSION SYSTEMS			

Figure 6.2   Sample energy source evaluation checklist.

PRELIMINARY HAZARD LIST

PROGRAM: _____   DATE: _____

ENGINEER: _____   PAGE: _____

ITEM	HAZARDOUS CONDITION	CAUSE	EFFECTS	RAC	COMMENTS
Assigned Number Sequence	List the nature of the condition (refer to Generic Hazard Group, If necessary).	Describe what is CAUSING the stated condition to exist	If allowed to go uncorrected, what will be the effect or effects of the hazardous condition?	Hazard Level assigned	Provide supporting comments and/or descriptions of rationale used to form conclusions

Figure 6.3   *Sample preliminary hazard list (PHL) worksheet.*

PRELIMINARY HAZARD ANALYSIS

PROGRAM: _____   DATE: _____

ENGINEER: _____   PAGE: _____

ITEM	HAZARDOUS CONDITION	CAUSE	EFFECTS	RAC	ASSESSMENTS	RECOMMENDATIONS
Assigned Number Sequence	List the nature of the Condition (refer to Generic Hazard Group, Group, if necessary).	Describe what is CAUSING the stated condition to exist	If allowed to go uncorrected, what will be the effect or effects of the hazardous condition?	Hazard Level assigned	Probability or Possibility of occurrence: • Likelihood • Exposure • Magnitude	Recommended actions to eliminate or control the hazard NOTE: Use the Hazard Reduction Precedence Sequence

Figure 6.4   *Sample preliminary hazard analysis (PHA) worksheet* (Note: *Worksheet will provide for hazards identification, evaluation, and resolution*).

PRELIMINARY HAZARD MATRIX				DATE: _____		
SYSTEM/PROGRAM EVALUATED:_____			PERFORMED BY: _____			

HAZARD GROUP	POTENTIAL FAILURE AREAS					
	Structural	Electrical	Pressure	Leakage/ Spill	Mechanical	Procedural
SLIPS, TRIPS, FALLS						
FALL FROM HEIGHT						
DECOMPOSITION						
ELECTRICITY						
FIRE / EXPLOSION						
ASPHYXIATION						
PATHOGENS						
MENTAL DISORDERS						
TEMPERATURE VARIATIONS						
RADIATION / LASERS / UV LIGHT						
MOVING EQUIPMENT / PINCH POINTS						

Figure 6.5   Sample preliminary hazard matrix (Note: *Examples have been used under "Potential Areas For Failure"*).

Contamination: The release of toxic, flammable, corrosive, condensable, or particulate matter into a system. Typically caused by leakage, spillage, loose objects, abrasion, growth, or component failure.

Corrosion: Structural degradation of metallic or nonmetallic equipment. Can be caused by leakage of reactive material, material incompatibility, or environmental conditions.

TABLE 6.1   Sample Generic Hazard Groups

Generic hazard groups

Slips, trips, falls
Fall from height
Decomposition
Electricity
Fire/explosion
Asphyxiation
Pathogens
Mental disorders
Temperature variations
Radiation/lasers/UV light
Moving equipment
Pinch points

Electrical Shock: Personnel injury or fatality due to electric current passing through any portion of the body. Typically caused by contact with energized electrical circuit, procedural error, component failures, static discharge, human factors, or environmental conditions. Can also degrade equipment operation.

Fire: Rapid oxidation of combustibles. Caused when fuel and oxidizer are exposed to an ignition source. Hypergolic fuels also ignite without an external source of ignition. Typically caused when fuels are raised above their ignition temperatures in the presence of an oxidizer or heat source.

Explosion: A violent release of energy due to overpressurization. Can be caused by fire, chemical reaction, excessive temperature, component failure, or procedural error.

Loss of Habitable Atmosphere: Removal or displacement of oxygen to below 19.5% by volume. Can be caused by a wide variety of conditions and situations.

Pathological: Injury to persons caused by disease, bacteria, microorganisms, and so on.

Psychological: Injury to persons due to mental conflicts such as sudden noises, perceived danger, preoccupation, or distraction, and so on.

Temperature Extremes: Injury to persons or damage to equipment due to departure of temperature from normal range. Extreme heat or cold can be due to the introduction of fire or cryogenics, respectively. May also be caused by component failure or procedural error. Results can be burns and/or structural damage.

Radiation: Exposure of persons or sensitive equipment to ionizing radiation, nonionizing radiation, ultraviolet or infrared light, lasers, electromagnetic or radio frequency emanations. Results can be burns to persons, structural damage to equipment, initiating of ordnance devices, and so on.

Figure 6.3 is a PHA worksheet. Upon comparison to Figure 6.4, the difference between the PHL and PHA worksheet is the *level of detail* in the PHA. While the PHL typically identifies the hazardous event, casual factors, system effects, and the RAC, the PHA goes further by providing recommended actions and referencing standards which are in violation, if applicable.

Fundamentally, there are basic questions which must be asked when developing the PHA. Although some may seem obvious or somewhat simplistic, they should still be considered. If such questions are not asked, the system safety professional runs the risk of conducting an incomplete analysis. All too often, the most obvious or visible may tend to conceal some associated level of risk of exposure to a hazard(s). Basic questions which should be resolved include

- What is the process/system under analysis?
- Does it involve people?
- What must the system always do?

- What must the system never do?
- Are there any codes/standards to address this?
- Has the system been used before?
- What does the system produce?
- What elements are taken into the system?
- What elements are discharged from the system?
- What could cause a release of a hazard?
- What is the assessment of this release?
- What/where are the energy sources/barriers?
- Is timing critical to safe operations?
- What are the inherent generic hazards in the system?
- How could control be improved?
- Will management accept these controls?

Tools such as the Hazard Element Evaluation Checklist, Figure 6.1, and the Energy Source Evaluation Checklist, Figure 6.2 (TAI 1989), can be used to facilitate the development of the PHL, Figure 6.3. The PHL, together with the PHA matrix can then be utilized to prepare the PHA worksheet.

The PHA matrix, Figure 6.5, and the PHA worksheet, Figure 6.4, both serve to formalize the question/answer process associated with PHA development.

The PHA Report

Once all the data have been evaluated and the PHA worksheet is completed, a formal report should be written documenting the results of the analysis. The narrative report typically includes a summary of all significant findings associated with operational risk. Recommendations for hazard elimination/control are also included in the report as well as suggestions for follow-on analyses. Although not entirely necessary, depending on the nature of the operation, process, or system, it is also useful to include a brief description of the project itself, its purpose and/or function as it relates to overall operations. The PHA/PHL worksheets are usually provided in the report as backup data to verify the report contents. Finally, the PHA report should also include a brief discussion of the methods used to develop the analysis (ETBA, FMEA, checklists, matrices, etc.), so that the reader can validate the report data, if required (Stephenson 1991).

PHA EXAMPLE

In an effort to demonstrate the utility of the *preliminary hazard list* and the *preliminary hazard analysis* in the initial evaluation of system risk, an example of a simple vapor degreaser in a manufacturing facility will be examined. This illustration will utilize the PHL in the development of the PHA in the method discussed earlier in this chapter. However, it must be noted from the onset that this example is intended to aid in the understanding and use of the basic PHL and PHA concepts. It is therefore only an example and cannot, in the interest of space and feasibility, be considered an *all-encompassing* preliminary hazard analysis of a vapor degreasing system. Therefore, this analysis will *only highlight* some of the many possible hazard risks associated with this proposed project.

System Description

Figures 6.6 and 6.7 show the basic concept and layout of the proposed vapor degreasing system. It will be comprised of a single, open-top solvent degreasing tank, contained in a concrete pit. The proposed tank measurements are 6 feet long, 4 feet wide, and 4 feet deep. The pit measurements are only slightly larger than the tank. The tank is to be constructed of steel with welded seams. Inside the tank, approximately 20 inches from the top, a 1-inch wide cooling coil is built into the wall of the steel tank structure. This coil, which contains a super-cold refrigerant, travels around the entire inner circumference of the degreasing tank. The refrigerant source is a 2-gallon reservoir tank located adjacent to the tank-pit area. A small electric motor is used to circulate the coolant through the coil. A compressor is also built into this system to ensure the refrigerant remains at a designated temperature. In order to generate the desired degreasing vapors, the solvent in the tank is heated to approximately 118–120°F, at which point it becomes a gas. As the heated gas moves toward the cold air near the cooling coil, it instantly vaporizes into a cloud which remains permanently

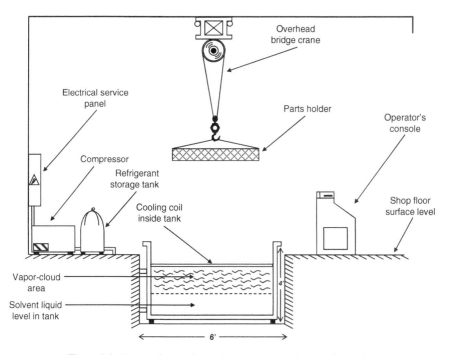

Figure 6.6   *Proposed vapor degreasing operation work area: planar view.*

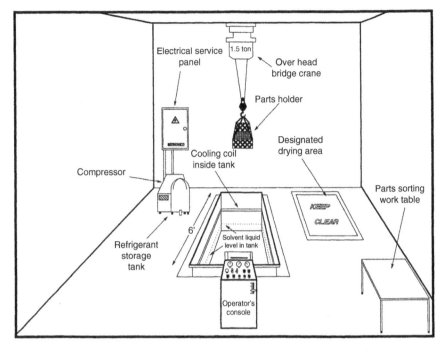

Figure 6.7   Proposed vapor degreasing operation work area: dimensional view.

suspended inside the tank, as long as the tank heat is applied or the coil remains cool. This vapor-cloud blanket, which will generally encompass a 10- to 15-inch thick space inside the tank, is where the degreasing will take place.

1,1,2 trichloro-1,2,2 trifluoroethane (Freon R 113) has been selected as the solvent to be used as the degreasing agent because of its relatively high toxicity threshold (1000 ppm), its nonflammable, noncombustible characteristics, as well as its excellent degreasing capabilities. The vapor degreasing tank is designed to hold up to 120 gallons of liquid. However, the operating procedures for this system specify a maximum of 100 gallons of liquid at any given time.

System Operation

The vapor degreaser is to be used to clean finely machined metal parts prior to painting. The parts will be placed in a specially designed holding rack and lowered into the vapor tank utilizing a small crane, which has been installed over the degreasing work station. As the parts reach the vapor cloud inside the tank, the operator is required to stop the crane and monitor a timer. The parts should remain in the vapor cloud for not more than 10 minutes, otherwise metal-pitting might occur and the relatively expensive parts would be ruined. After the parts are cleaned, they are removed and placed in a designated drying area where they are air-dried for 15 minutes, inspected, and removed for painting.

Preliminary Assessment

Since we have begun this analysis during the concept phase of the project life cycle, it would be prudent to develop a preliminary hazard list of basic safety concerns associated with the project concept. Then, as the project moves into the design phase, a PHA can be performed based upon the information contained on the PHL.

In considering the limited amount of information provided above regarding the proposed project, several questions can now be asked to determine the level of possible hazard risk associated with system operation. Utilizing the Generic Hazard Evaluation Checklist (Figure 6.1) and the Generic Source Evaluation Checklist (Figure 6.2), the analyst can begin to identify and address the fundamental inputs required to develop the PHL worksheet (Figure 6.3). Once each checklist is complete, the information can literally be *transferred* to the PHA worksheet for further analysis and development. Figure 6.8 is a completed Hazard Evaluation Checklist for the vapor degreasing system. Note the utility of the checklist and its ability to assist the analyst in isolating hazard concerns, as well as eliminating those areas which are not applicable. Already, in these early stages of analysis, the system safety effort is beginning to focus on essential hazard concerns that will eventually be fully addressed, evaluated, controlled, and/or eliminated. Figure 6.9 uses the Energy Source Evaluation Checklist for the proposed degreasing project. The PHL can now be completed in greater detail using the data provided here.

Evaluation of System Risk

The information provided above on this proposed system reveals many serious or potentially serious hazard risk levels. When asking the basic questions associated with the identification of system risk, the analyst can begin to categorize the severity of a potential mishap and evaluate the probability of a possible occurrence (refer Tables 2.1 and 2.2). The following is an itemized listing of a few of the initial safety concerns which should be resolved prior to proceeding into the design phase of this project's life cycle. The identification of these potential hazard risks is the result of proper utilization of the basic system safety tools discussed thus far.

- Crane Operations

   Hazard Risk Item 1: Personnel under suspended loads

   The proposed design will require that the crane/load combination be moved directly over the operator's console. OSHA requires that crane operators avoid transporting loads over personnel [OSHA 29 CFR §1910.179(n)(3)(vi)].

   Risk Assessment 1: 2A (refer Tables 2.1–2.3)

   Because the severity of such an occurrence is categorized as "critical" (i.e., severe injury, occupational illness, or system damage) and, due to the proposed system design, the probability of a mishap of this nature is categorized as "frequent" (i.e., likely to occur frequently), the risk assessment matrix

GENERIC HAZARD EVALUATION CHECKLIST

SYSTEM/PROGRAM: Vapor Degreaser

PERFORMED BY: J. Doe   **DATE:**   05/2013

HAZARDOUS ELEMENT SOURCES	YES	NO	COMMENTS
ACCELERATION	✓		Crane movement into/out of tank; possible movement over operator
CHEMICAL (DISSOCIATION, REPLACEMENT, SUBSTITUTION)	✓		Use of Freon 113 violates provisions of the Clean Air Act of 1990 (CFC Phase-out)
ELECTRICAL ELEMENTS AND OPERATIONS	✓		Heating elements/controls in tank to heat Freon 113
ENVIRONMENT	✓		Closed room, no ventilation provisions; Pit is open-top with no barriers installed
LEAKAGE	✓		No provisions for tank leakage into pit (Containment? Emergency Response? etc.)
MOISTURE	✓		Condensation on outside surface of steel tank around cooling coil
OXIDATION	✓		Possible rust/corrosion problems due to moisture build-up on steel tank (see above)
OFF-GASSING OF MATERIAL PROPERTIES	✓		Heated Freon will off-gas. Hydrofluoric acid gas is one hazardous by-product
PRESSURE HIGH, LOW, RAPID CHANGE		✓	N/A
STRESS	✓		Tank weld points –stress of liquid mass and temperature extremes
STRUCTURAL FAILURE	✓		Crane failure; tank failure

Figure 6.8   Vapor degreaser: generic hazard evaluation checklist.

GENERIC ENERGY SOURCE EVALUATION CHECKLIST

SYSTEM/PROGRAM: Vapor Degreaser

PERFORMED BY: J.Doe   DATE: 05/2013

ENERGY SOURCE (Kinetic/Potential)	YES	NO	COMMENTS
ACTUATING DEVICES	✓		Crane system
CATAPULTED OBJECTS		✓	N/A
CHARGED ELECTRICAL CAPACITORS	✓		Heating mechanism; operator's control panel; crane system
CHEMICAL REACTION	✓		Potential Freon reaction with steel tank
CRYOGENIC MATERIAL	✓		Refrigerant in cooling coil
ELECTRICAL GENERATORS		✓	N/A
ELECTROMAGNETIC, IONIZING NONIONIZING RADIATION		✓	N/A
EXPLOSIVE CHARGES		✓	N/A
ELECTROSTATICE CHARGE OR DISCHARGE		✓	N/A
FALLING OBJECTS	✓		Crane securing, proper rigging; structural integrity of crane
FUEL/PROPELLANTS		✓	N/A
GAS GENERATORS		✓	N/A
HEATING DEVICES	✓		Heating element in tank
INITIATORS/IGNITORS		✓	N/A
NUCLEAR		✓	N/A
PLACEMENT OF SYSTEMS, COMPONENTS	✓		Tank open pit, no barriers or controls; Crane movement over operator's head
PRESSURE CONTAINERS		✓	N/A
PUMPS, BLOWERS, FANS		✓	N/A
ROTATING/MOVING MACHINERY	✓		Crane system
SPRING – LOADED DEVICES		✓	N/A
STORAGE BATTERIES		✓	N/A
SUSPENSION SYSTEMS		✓	N/A

Figure 6.9 *Vapor degreaser: generic energy source evaluation checklist.*

assigns this risk a classification of 2A which, according to Table 2.3, is an unacceptable level of risk.

Recommendation 1: To ensure maximum possible reduction of the hazard risk associated with personnel working under suspended loads, the design should consider either relocation of the operator console, or, installation of a positive stop in the crane trolley to prevent any movement beyond a predetermined point.

Hazard Risk Item 2: Liquid Solvent Contacting Parts

The crane operator is required to lower the parts holding rack into the vapor degreasing tank to a level which places the parts directly in the center of the vapor cloud blanket. This effort requires visual contact between the operator and the inside of the vapor tank. From the control console, it may not always be possible for the operator to have total visual contact at all times with the parts rack, especially when it enters the tank. If the parts are lowered too far, they could end up submerged in the liquid solvent and possibly be ruined. Also, if there is a failure of the crane wire rope or the sling supporting the parts rack, the parts might fall into the liquid and retrieval would be extremely difficult.

Risk Assessment 2: 2B (refer Tables 2.1–2.3)

Liquid contact with parts is assessed as a "critical" occurrence, since the potential damage to the parts would most likely render them unusable. The likelihood of such a mishap is considered highly "probable," based on the proposed system design. The Risk Assessment Matrix (Table 2.3) indicates that a risk classification of 2B is unacceptable. Therefore, the system safety precedence tells us that such risk should be approached with the intention of elimination, or possible reduction to an acceptable level.

Recommendation 2: To prevent the potential for operator error, the design should provide another automatic stop for the crane so that the parts cannot possibly be lowered any further into the tank than the required level. Also, as an added precaution or as a possible alternative to the automatic crane stop, a raised mesh floor could be installed in the tank just above the liquid level so that contact with the solvent liquid would not be possible in any case.

Hazard Risk Item 3: Parts Remain Too Long in Vapor

The procedure for this operation indicates that if the machined parts remain in the vapor cloud for more than a 10-minute period, possible pitting will occur and the expensive parts will be ruined. To prevent this occurrence, the operator will be required to monitor a timer and remove the parts rack when the 10 minutes has elapsed. It is estimated that the degreasing operation will occur approximately 25 times during an 8-hour shift. Based upon this frequency and the monotony associated with any clock-watching activity, the probability of an operator error is considered extremely high.

Risk Assessment 3: 2C (refer Tables 2.1–2.3)

Because the probability of the above described mishap has been assessed as "occasional" (i.e., likely to occur sometime in the life of an item) and the severity of the mishap outcome can be categorized as "critical," the risk classification from Table 2.3 indicates that such a mishap is assessed as 2C. This level of risk, although not totally unacceptable, by definition, is still undesirable and must therefore be reduced or eliminated through design controls.

Recommendation 3: One way to eliminate the risk of this occurrence is to automate the system so that, once the operator lowers the crane into the tank and the crane stops at the predetermined level, the crane will automatically begin to rise out of the vapor cloud after the 10-minute degreasing operation. However, while such a system would eliminate the risk, it would, of course, be very expensive to install and maintain. There are also inherent hazards that exist whenever any equipment starts automatically. As an alternative, installation of a simple automatic timing device in the operator's console that will sound an alarm after the 10-minute period has elapsed will alert the operator to remove the parts from the tank. This alternative recommendation would not totally eliminate the risk, but it will reduce, or control it to a much lower and therefore acceptable level.

- Tank Design/Structural Considerations

Hazard Risk 4: Welded Seams on Tank

The welded seams on the vapor degreasing tank offer a potential hazard risk. Failure of any weld point below the liquid level would obviously result in loss of some or all of the liquid solvent into the surrounding concrete pit area. Considering the stress of liquid mass inside the tank, as well as the temperature extremes associated with solvent heating and off-gas cooling, the possibility of weld failure occurring sometime during the life cycle of this system must be carefully evaluated.

Risk Assessment 4: 3D (refer Tables 2.1–2.3)

Although weld seam failure can occur, its likelihood can be described as "remote" (i.e., unlikely, but may possibly occur in the life of an item). If the seams were to fail, the severity of the outcome is assessed as "marginal" (i.e., minor injury, occupational illness, or system damage). Hence, the Risk Assessment Matrix classifies this incident as acceptable, with a review by engineering and management personnel.

Recommendation 4: Although such a risk may be considered acceptable, the system safety analyst is still obligated to recommend possible risk reduction techniques to provide management with additional criteria by which to accept the level of risk. In this case, one possible recommendation would be to install a unibody tank (i.e., all one piece, custom designed, with no welded seams). This would totally eliminate the risk of tank failure due to faulty welded seams. However, because such a system would most likely be extremely expensive, management is not likely to consider a unibody tank design. A more economic alternative would be to X-ray each welded seam prior to tank installation. The X-ray will show any faulty or improperly welded seams which could then be corrected. Although the risk of weld seam failure will not be entirely eliminated through the use of X-ray analysis, the risk of failure would be greatly reduced to the point of highly improbable and, therefore, acceptable.

Hazard Risk 5: Tank Layout and Design

This particular design provides no means of accessing the exterior wall of the tank for inspection, once it is installed in the pit. The close design tolerances between the exterior tank wall and the concrete pit will not permit maintenance or inspection of such possible failure components as the solvent-fill connection point, the heating element and associated circuitry, the refrigerant connection to the cooling coil, and general tank structural integrity. Also, in the unlikely event of tank leakage, or the possible release of Freon liquid from the solvent-fill connection point, the solvent would accumulate inside the pit area. Depending upon the amount of the release, there might not be sufficient containment space in the pit. Removal of spilled solvent would also be difficult because of the close design tolerances.

Risk Assessment 5: 4D (refer Tables 2.1–2.3)

The probability of tank leakage might be assessed as "remote," especially if X-ray analysis of welded seams is performed prior to installation. Even if it were to leak, the resulting severity would be considered negligible (i.e., less than minor injury, occupational illness, or system damage) since the liquid would be contained in the concrete pit area. Therefore, the Risk Assessment Matrix recommends such a risk be classified as acceptable, without review.

Recommendation 5: Even though the risk may be classified as acceptable without review, the system safety effort cannot take for granted any level of risk without providing management with possible alternative design recommendations for their consideration and review. As stated above, there is a slight possibility that the pit might not be able to contain all the liquid in the event of a major leak or spill. Also, the inspection and maintenance of tank system components is not entirely possible based upon the proposed design. The concrete pit could be designed with much larger dimensions than that of the tank. This design change would permit authorized access into the pit area for inspection and maintenance of the tank structure and any of its components or connections. Also, if a pit redesign is considered, the new layout should allow for total liquid containment. This could be accomplished by constructing a sump area in the pit and installing a raised, corrugated floor. The tank would sit on the suspended steel floor instead of the pit bottom itself. If a leak were to occur, it would all be contained below the base of the tank. A sump pump could be installed in the pit that would automatically activate and remove any liquid from the sump area. If an automatic pump proves too cost prohibitive, a manual pump could be lowered into the containment area to transfer the liquid out of the sump and into drums for disposition and removal as a hazardous waste. Of course all these recommended design changes will obviously increase the cost of the entire project and must therefore be carefully considered. The point of the system safety effort is to provide management with risk reduction choices so that accurate informed decisions can be made.

- Electrical: Design and Layout Considerations

Hazard Risk 6: Location of Service Panel

The electrical service panel in the vapor degreasing work area is located on the other side of the tank pit, away from the operator's console and behind a compressor and the refrigerant storage tank. Aside from violating OSHA criteria regarding accessibility of electrical service panels [OSHA 29 CFR §1910.303(h)(3)], the proposed design does not afford quick access to circuit breakers in the event of an emergency. If a problem occurs with the crane, the compressor, or the solvent heating system, and the operator is unable to quickly de-energize the electrical power source, then the possible risk of hazard might be quite severe, depending upon the nature and extent of any such problems.

Risk Assessment 6: 2C (refer Tables 2.1–2.3)

Obviously, such "critical" mishaps as those described above might possibly occur during the life cycle of this system (occasional). The Risk Assessment Matrix indicates that such risk is undesirable and must, if at all possible, be controlled/reduced to acceptable levels or eliminated totally.

Recommendation 6: A redesign of the layout is required to ensure appropriate access to the electrical service panel. If possible, it should be located adjacent to the operator console area so the operator can easily access the electrical circuits in the event of an emergency. If total relocation of the service panel from one side of the room to another is not possible or feasible due to building limitations, room design, and so on, then the operator's console should be provided with a "dead-man" switch that will de-energize all electrical power to the vapor degreasing system, including the crane. This relatively simple solution will provide the operator more control over the risk associated with critical electrical systems and subsystems.

Other possible areas of consideration for hazard risk include, but are certainly not limited to the following:

- Pit/Tank Area: Wide-open, no barricades, personnel could fall into tank. Recommend barriers such as stanchions with rope or chain, or guard rails as required by OSHA at 29 CFR §1910.23(a)(5).

- Parts Holder Synthetic Web Sling: Use of synthetic web slings in areas where fumes or vapors are present is a violation of OSHA 29 CFR §1910.184(i)(6). Constant exposure to solvent vapors could result in premature failure of synthetic web sling. Recommend use of wire rope sling instead.

- Room Ventilation: A remote potential does exists for oxygen displacement in the room if the cooling coil inside the tank should fail. Such a failure would allow Freon gas to accumulate in the work area and deplete the level of oxygen below the life sustaining level of 19.5% by volume. Recommend installation of an exhaust ventilation system as required by applicable regulations and/or manufacturer recommendations or requirements.

- CFC Phase-Out Requirements: The solvent selected for use in the degreasing tank operation is listed as a designated chlorofluorocarbon, the use of which was subjected to the *phased-out* requirements stipulated under the Clean Air Act of 1990. Since this system is still in the design phase, it is not too late to consider an alternate degreasing solvent that is not a threat to the earth's ozone layer. *(NOTE: While it is true that the use of Freon 113 is not a particular hazard to personnel or the equipment used in this degreasing process, its use still poses an alleged threat to the earth's environment and, therefore, its identification as a potential "hazard" is warranted).*

The list of potential hazard risk will continue to grow, the more extensive the analysis becomes. There are many other possible design hazards associated with this system which must be considered and evaluated during the concept phase of the project life cycle. However, for the purpose of this example, the above listed items have been provided and discussed to demonstrate the level of analysis which is typically conducted when developing a PHL. Figure 6.10 is an example of how some of these identified hazards might be recorded on the PHL worksheet. Please note that in this example only five of the many possible hazards have been recorded. As the preliminary hazard list is finalized, much of the information pertaining to identified hazards that have not been corrected or controlled can be transferred to the PHA worksheet as the project enters the design (as shown on Figure 6.11). Finally,

PRELIMINARY HAZARD LIST					
PROGRAM: Vapor Degreaser				DATE: 05-06-2013	
ENGINEER: Jane Doe				PAGE: 1 of 1	
ITEM	HAZARDOUS CONDITION	CAUSE	EFFECTS	RAC	COMMENTS
1	COLLISION	Structural failure of crane equipment	Injury to personnel; Damage to tank or equipment	2A	RE: ANSI B30 requirements for cranes
2	COLLISION	Crane lowers parts too far into tank, submerging parts in liquid Freon	Possible loss of usable parts	2B	No provision to prevent the improper use or operation of cranes
3	STRUCTURAL	Tank leakage or structural failure	Injury to personnel; loss or damage to parts or equipment	3D	Weld points are not inspected before initial use
4	TEMPERATURE EXTREMES	Solvent is heated to 120°F; Cold liquid refrigerant also present	Possible personnel injury upon contact with either	4D	Contact not probable due to isolation of chemicals
5	LOSS OF HABITABLE ATMOSPHERE	Cooling-coil system failure during tank operations	Possible personnel injury or death due to displacement of oxygen in room	2D	RE: ANSI codes for ventilation requirements
6	ELECTRICAL SHOCK	Short in heating system	Possible personnel injury or death	2A	RE: NEC and OSHA requiirements for electrical safety

Figure 6.10   Vapor degreaser: preliminary hazard list worksheet.

PRELIMINARY HAZARD ANALYSIS

PROGRAM: Vapor Degreaser

ENGINEER: Jane Doe

DATE: 05-05-2013

PAGE: 1 of 1

ITEM	HAZARDOUS CONDITION	CAUSE	EFFECTS	RAC	ASSESSMENTS	RECOMMENDATIONS
1	COLLISION	Crane Failure	Injury to personnel; damage to tank or equipment	2A	Personnel possibly working under suspended loads	1. Install positive stop mechanism in crane path to avoid exposure 2. Relocate Operator's Console away from crane movement path
2	COLLISION	Crane Failure	Parts damaged due to contact with liquid solvent	2B	Unacceptable risk due to high probability of liquid contact	1. Install positive stop mechanism in crane path to prevent possibility of liquid contact 2. Install suspended floor inside tank, above liquid level
3	STRUCTURAL	Tank Weld Failure	Solvent leak into outer pit area; system damage	3D	Remote possibility due to stress and temperature extremes	1. Install uni-body tank 2. X-ray welded seams prior to installation

Figure 6.11 Vapor degreaser: preliminary hazard analysis worksheet.

the PHA *Report* can be generated based upon the evaluation and analysis of system hazard risk, as described earlier in this chapter.

SUMMARY

In system safety analysis, the initial process begins with the development of the *preliminary hazard list* during the project or system *concept phase*. Although it is not always compiled in all cases, an available PHL can become the working foundation for the development of the *preliminary hazard analysis* during the *design* phase of the project life cycle.

Use of a variety of system safety concepts and tools, such as the *order of precedence* for hazard reduction, the hazard *severity* and *probability* tables, and the *hazard risk matrix*, will assist the analyst in determining the appropriate *risk assessment code* to assign to a particular hazard risk. The RAC will prioritize for management the specific level of risk associated with a specific, identified hazard concern.

The information recorded on the PHA worksheet together with the PHA report, will greatly facilitate the performance of other beneficial system analyses (such as the Subsystem Hazard Analysis, the Failure Mode and Effect Analysis, and the Operating and Support Hazard Analysis) which may be accomplished during the remaining phases of the product life cycle.

7

Subsystem and System Hazard Analyses

INTRODUCTION

A *Subsystem Hazard Analysis (SSHA)* or a *System Hazard Analysis (SHA)* may be required depending upon the complexity of a given program or project. Both the SSHA and the SHA are often referred to as one-in-the-same by many system safety professionals (Stephenson 1991). However, as explained here, the two methods are slightly different and, if used properly, they provide for a more complete evaluation of a given system.

The SSHA can provide specific details about the hazards associated with many subsystems, or only one subsystem. The SHA provides an analysis of the system as a whole, with all subsystems working together. The SSHA should be performed as early in the design phase as possible. Some experts recommend the initial SSHA be conducted as early as 35% of design completion. However, realistically, exact timing of the SSHA will usually be dependent upon the availability of the required subsystem data such as project description documentation, complete and accurate drawings and schematics, and applicable regulatory codes and/or safety design standards. The PHL (if available) and the preliminary hazard analysis (PHA) should also be used in preparing the SSHA. Any relative lessons learned will also be effective in developing the SSHA. As a minimum, the SSHA should address at least the following elements of the hazard identification process:

- Hazard-initiating component(s)
- Component hazardous modes

Basic Guide to System Safety, Third Edition. Jeffrey W. Vincoli.
© 2014 John Wiley & Sons, Inc. Published 2014 by John Wiley & Sons, Inc.

SYSTEM/SUBSYSTEM HAZARD ANALYSIS

SYSTEM/SUBSYSTEM: _____

PROGRAM: _____   DATE: _____

ENGINEER: _____   PAGE: _____

ITEM	HAZARDOUS CONDITION	CAUSE	EFFECTS	RAC	RECOMMENDED CONTROLS	CONTROLLED RAC	STANDARDS

Figure 7.1   Sample subsystem/system hazard analysis (SSHA) worksheet.

- System operational mode(s) for each component
- Hazard effects of each operational mode

Once all the available data have been examined, the system safety engineer can then begin to develop the SSHA using the SSHA worksheet, as shown in Figure 7.1. It should be noted again that the SSHA is intended to provide an analysis of the hazards associated with an individual subsystem or group of systems. Therefore, it should also be emphasized that the SSHA, by itself, will not provide a complete picture of the entire system or any hazards not related to the specific subsystems examined. Although this principle is somewhat basic, it must be clearly understood prior to initiating any SSHA. There is a danger of overlooking other hazards of the total system if the SSHA is taken alone without adequate consideration of the entire project. To prevent the potential for this occurrence, the SHA should be developed using data obtained during the performance of the SSHA. To conduct an SSHA, it is recommended that the system safety engineer utilize the Failure Mode and Effect Analysis (FMEA) and/or the Fault Tree Analysis (FTA) techniques, as described in Chapter 10 and Chapter 12, respectively, in this text.

The Subsystem Hazard Analysis Report

The SSHA report is completed after the SSHA is fully developed. If more than one SSHA is performed (i.e., at the 35%, 50%, and/or 65% design phase), a detailed report should be prepared after each SSHA to adequately document any findings and recommendations (Stephenson 1991). The SSHA report will typically contain a full description of the subsystem(s) which were the subject of the analysis and their relationship/function with regard to total system operation. A summary of all findings resulting from the analysis are also provided in the narrative of the report, as are any recommendations to improve hazard control and risk reduction. The report should also provide an evaluation of existing or planned hazard controls as well as identify areas where further controls are recommended or required. The SSHA will further clarify the control of any risk associated with a hazard by assigning a *"controlled RAC"* to the hazard, after recommended controls have been put into place. The use of the controlled RAC facilitates the accurate tracking of a specific hazardous condition as it relates to a specific subsystem. As with the PHA report, the SSHA report should discuss the techniques and methodology utilized during the performance of the SSHA and provide copies of the SSHA worksheets used to develop and finalize the report. Finally, any criteria for risk acceptance should also be detailed in the report to establish the baseline from which the SSHA had been developed. Such criteria may have been developed by a variety of means such as customer preference or past performance experience.

SSHA EXAMPLE

To understand the utility of the SSHA, an example of a hydraulic elevator will be evaluated. A typical hydraulic elevator system consists of many subsystems such as car buffers and bumpers, plunger mechanism, counterweight system, car door system, car safeties, counterweight safeties, and cylinders. For simplicity, this SSHA will consider only the hydraulic plunger subsystem. This SSHA is being performed on an existing elevator (i.e., during the *operational* phase).

The elevator is one of two that serve an extremely busy three-story office building. The SSHA is necessary as a result of two individual reports from passengers claiming a "slipping sensation" in the car while waiting for the doors to open at their designated floor landing. When the doors finally opened, the normally level car was uneven by as much as 4 inches. An incident investigation has been ordered by the building operator to determine possible causes of the alleged slippage. As part of the problem assessment process, the SSHA will evaluate the plunger system to determine potential hazards, identify the possible cause and effect of those hazards, and to establish a risk assessment code which will categorize the level of acceptable risk associated with any identified hazard. These findings will be recorded on the SSHA worksheet. Once all hazard conditions are understood, the SSHA will provide recommended control actions. The analyst will also show, on the SSHA worksheet, a modified risk assessment code that reflects the level of acceptable risk each hazard would be reduced to, once recommended control actions are in place.

System Description

Figure 7.2 is a diagram of the major hydraulic elevator components, showing the plunger location in relation to the other associated subsystems. As a subsystem, the

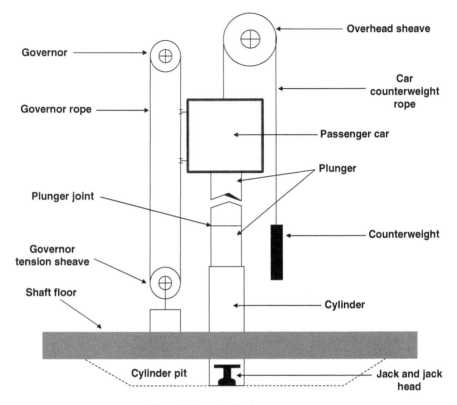

Figure 7.2   *Hydraulic elevator system.*

plunger mechanism consists of two primary components that will be evaluated in this SSHA: the plunger joints and the hydraulic cylinder.

The plunger joint is a connection in the plunger extension mechanism that permits the telescoping action during operation which causes the desired lifting or lowering movement of the elevator car. There are a total of two plunger joints in this system. Figure 7.3 shows the components of the plunger joint.

The hydraulic cylinder contains the hydraulic oil necessary to enable the lifting action of the elevator. The total vertical travel distance of this particular elevator is 26 feet (7.92 meters) and the cylinder volume is 3 gallons per foot (11.4 liters per meter). A total of 77 gallons (291.5 liters) of oil is encased by the cylinder when the elevator is in the uppermost position in the shaft. When fully lowered, the volume of oil in the cylinder is decreased to 42 gallons (159 liters). There is another 15 gallons (58.6 liters) of oil reserve over and above that required to raise the car to its maximum extension. There is an additional 4 gallons (15.6 liters) of oil contained on the lines connecting the power unit to the cylinder. Hence, the total oil contained in this system during full extension is equal to 96 gallons (383.4 liters). The hydraulic oil was filled at the time of installation, and the same oil is expected to serve the

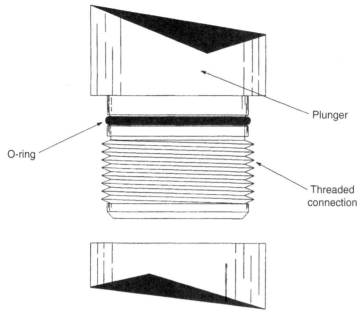

Figure 7.3 *Plunger joint.*

system throughout the entire operational phase of the product life cycle (estimated at approximately 40 years). The lifting jack, located inside the cylinder, operates under hydraulic pressure to move the plunger up or down.

Evaluation of Subsystem Hazard Risk

The information provided above on the elevator plunger subsystem reveals some component areas that will require further examination. The intent of this particular SSHA is to identify possible hazardous conditions caused by a failure of one or more components in the plunger subsystem that could possibly result in the reported slippage of the elevator car in the shaft. Once these hazards are identified and the cause and effect determined, the SSHA can recommend control actions to reduce the risk to a lower, more acceptable level. In analyzing the elevator to the subsystem level, it would be prudent to review any previous analyses that may have been performed on this system during the design or production phases. In particular, the PHA will offer excellent, fundamental risk assessment information. Of more utility in this case would be the data recorded on any failure mode and effect analyses performed on the elevator system. Depending on the nature and scope of the FMEA, it may provide precise information pertaining to the plunger subsystem in which case the subsequent SSHA process will be greatly facilitated. For the purpose of this example, we will assume that the previous FMEA on the elevator system did not examine the plunger components in particular and, therefore, no prior hazard analysis information is available for evaluation.

The SSHA has identified two possible conditions that could cause the passenger car to slip in the shaft once it has come to a stop at a designated floor landing.

- **Plunger Joint**
  Hazard Condition Item 1: Loss of Hydraulic Pressure

  A loss in oil pressure due to a small system leak occurring around the plunger joint due to o-ring failure could result in a possible slipping of the elevator car until replacement oil is obtained from the reservoir or until the suspect joint was once again contained (submerged) within the plunger mechanism (i.e., as in retraction during car lowering). If the condition is allowed to continue, it is possible that the leak could become more serious with the resultant effect of pressure loss during operation and subsequent uncontrollable and undesirable lowering of the passenger car.

  Risk Assessment Code, Item 1: 3D

  The required design safety factors and hydraulic pressure bleed-off rates will not permit a catastrophic result in the event of pressure loss. At most, in the event of a continuous oil pressure loss, the elevator would descend slowly to the ground floor level at a rate consistent with that of the loss. Therefore, a severity level of marginal has been assigned and a probability of occurrence level of remote has been determined.

  Recommendation 1: Even though an RAC of 3D is considered acceptable with review, hazard risk reduction is still warranted, especially in consideration of the two reported incidents. In order to ensure maximum possible reduction of the hazard risk associated with a failure of the plunger joint components, it is recommended that routine preventative maintenance inspections include replacement of the plunger joint o-ring component. This action would further reduce the probability of occurrence to the improbable level. Once recommended control actions are in place, a controlled RAC of 3E (marginal/improbable) can be assigned.

- **Hydraulic Cylinder**
  Hazard Condition Item 2: Loss of Hydraulic Pressure

  A small leak in the hydraulic cylinder system would result in a loss of pressure in the cylinder and a subsequent inability to maintain constant cylinder volume. Any damage to the cylinder jack head would cause a leak of hydraulic oil from the cylinder system. Such damage is possible due to an abrupt contact with the plunger mechanism and/or a failure in the neoprene seal between the jack head and the plunger. The system safety analyst has learned from the PHA report that there is no provision for oil recovery in the event of damage to the jack head.

Risk Assessment Code, Item 2: 2B

Loss of oil due to jack head damage is probable since no provision currently exists for oil recovery. This condition has been assessed as critical due to the potential result of such a hazard (i.e., total and sudden pressure loss in the cylinder system; possible rapid, although not uncontrolled, descent of the passenger car).

Recommendation 1: Installation of an oil recovery ring on the jack head, as required by ANSI A17.1, Rule 1302.3h (Safety Code for Elevators & Escalators) would significantly reduce the hazard risk associated with this condition. Frequent inspection of the neoprene seal for damage and routine engineering evaluation of clearance tolerance between the plunger and jack head will also ensure system integrity and reduce risk. A controlled RAC of 2E (critical/improbable) could be assigned to this condition once the recommended controls are implemented.

Figure 7.4 is the SSHA worksheet, completed to show the analysis of the two conditions discussed in this example. Additional evaluation can occur regarding the elevator system, with concentration on other possible hazardous conditions associated with other elevator subsystems and components.

SYSTEM/SUBSYSTEM HAZARD ANALYSIS

SYSTEM/SUBSYSTEM: ELEVATOR PLUNGER

PROGRAM: Maintenance   DATE: 11-16-2005

ENGINEER: John Doe   PAGE: 1 of 1

ITEM	HAZARDOUS CONDITION	CAUSE	EFFECTS	RAC	RECOMMENDED CONTROLS	CONTROLLED RAC	STANDARDS
1	LOSS OF PRESSURE	Oil leakage from plunger joint(s)	Elevator passenger car may "slip" in shaft; alarm to passengers	3D	Routine and scheduled inspections; frequent O-ring replacement	3E	ANSIA17.1 Rule1302.1
2	LOSS OF PRESSURE	Oil leakage from cylinder or damage to jack head	Sudden and abrupt but not uncontrolled slippage of elevator passenger car in shaft, alarming passengers	2B	Provide oil recovery ring for jack head; ensure proper maintenance and inspection of cylinder plunger system	2E	ANSIA17.1 Rule1302.1

Figure 7.4 *Elevator plunger subsystem/system hazard analysis worksheet.*

SUMMARY

The SSHA evaluates hazardous conditions, on the subsystem level, which may effect the safe operation of the entire system. In the performance of the SSHA, it is prudent to examine previous analyses that may have been performed such as the PHA and the FMEA. Ideally, the SSHA is conducted during the design phase and/or the production phase, as shown in Chapter 3, Figure 3.4. However, as discussed in the above example, an SSHA can also be done during the operation phase, as required, to assist in the identification of hazardous conditions and the analysis of specific subsystems and/or components. In the event of an actual accident/incident investigation, the completed SSHA can be used to assist in the development of an FTA by providing data on possible contributing fault factors located at the subsystem or component level.

8

Operating and Support Hazard Analysis

INTRODUCTION

The purpose of the Operating and Support Hazard Analysis (O&SHA), sometimes called the Operating Hazard Analysis (OHA), is to

1. Identify all hazards in the operation of a system that are inherently dangerous to personnel, or in which a human error could be hazardous to equipment or people, and

2. Provide recommended risk-reduction alternatives during all phases of tasks or operations that are controlled by written procedures (TAI 1989; Stephenson 1991).

Simply stated, the O&SHA encompasses an analytical review of the controlling documents to ensure hazard elimination or control and concentrates heavily on the performance of people (human factors and human behaviors) and their relationship to the hazards within the task. The focus is primarily upon the *maintenance* and *operation* of the system, rather than the system components themselves.

ERGONOMICS

The O&SHA examines those operating functions which may be inherently dangerous to personnel, or in which personnel error could be hazardous to equipment or people.

Basic Guide to System Safety, Third Edition. Jeffrey W. Vincoli.
© 2014 John Wiley & Sons, Inc. Published 2014 by John Wiley & Sons, Inc.

TABLE 8.1   Typical Human Errors Commonly Made by the Three Groups Having Impact on a System

Operator	► Omit required actions
	► Performance of nonrequired actions
	► Failure to recognize needed action
	► Improper (early, late, wrong) response
	► Poor communications
	► Maintenance error
Designer	► Lack of hazard awareness
	► Designing requirement for repetitious tasks
	► Requiring quick operator response for hazard recognition/resolution
	► Requiring operator to perform rapid, complex computations
	► Design sensor requirements outside human range
	► Requiring continuous operator attention
	► Requiring operator to work in poor environment
	► Requiring simultaneous physical activity and communication
	► Designing or providing improper tools
	► Providing inadequate or faulty written procedures
Manager	► Providing improper or inadequate training
	► Creating unrealistic production schedules
	► Assigning inexperienced personnel to complex tasks
	► Providing inadequate oversight or guidance during work operations

In fact, ergonomics (the scientific study of the interrelationship between people, their occupations, and their work environment) is a major element of the O&SHA.

Humans are involved in almost every facet of most system development and operations. The human interface, therefore, presents the most critical area of system safety analysis (DOE SSDC-2 1976). It is a widely known and often too readily accepted fact that human error is a major causal factor (either primary or contributory) in many mishaps. Table 8.1 lists common errors which often result from a breakdown of the human component (DOE SSDC-2 1976). The information presented in the table obviously indicates that the designer has a heavy responsibility for reducing the probability that a human error will cause additional problems with the operation of the system. The designer must consider all possible outcomes of a given human error and attempt to design the system such that the end result will be a fail-safe situation (i.e., the system fails in a safe, nonenergy transferring mode). Because certain elements of human behavior are somewhat predictable, the analyst should take such elements into consideration when evaluating a human interface system. For example, people will usually follow procedures that involve

1. Minimal mental and/or physical effort;
2. Reduction in time to complete a given task;
3. Elimination of discomfort;
4. Elimination of monotony or fatigue.

TABLE 8.2   Characteristics of Common Human Errors

Physical	► Anthropometric data/limitations
	► Controls and barriers: types, direction, blind operations, etc.
	► Equipment and hardware design inadequate
Work space requirements	► Position of operator relative to activities performed
	► Task requirements—adequate space to perform
	► Limitations to ingress into and egress out of work area
	► Adequate isolation from other tasks
	► Loads are too heavy
Environmental	► Illumination adequate to perform task
	► Atmospheric conditions
	► Noise levels in area
	► Motion—vibrations of (or in) working environment
	► Poor or inadequate ventilation
	► Temperature extremes
Limitations	► Prolonged concentration
	► Personal stress (physical or mental)
	► Inadequate rest
	► Illness
	► Boredom

Human error has been defined as an action that is inconsistent with or contrary to established behavioral patterns considered to be normal, or that differs from prescribed procedures that may or may not result in an adverse or unwanted event (TAI 1989). The common causes of human error can be divided into the following four categories and evaluated in an O&SHA (TAI 1989):

1. Physical
2. Work space requirements
3. Environmental
4. Limitations to human performance

Each of these categories are further broken down in Table 8.2 for clarification. Of course these causal breakdowns are not all-inclusive and more can certainly be added to each category. The point is, once these causes are examined and determined applicable to a specific task or operation, the analyst can then recommend design principles and control methods to prevent occurrence or reduce the possibility of human errors.

WHEN TO PERFORM THE O&SHA

Because of the human factor element and the necessity to reduce risk of injury to personnel, the O&SHA should be performed as early in the product life cycle as possible, or, at the very least, prior to the first operational use of the system. This is not always entirely feasible because the design of the product or system must usually be practically complete before maintenance and operating procedures are developed. At any rate, the O&SHA should be complete at the end of the production phase, prior to first use (Stephenson 1991).

Exact timing of the O&SHA will obviously be contingent upon the desired end use of the product or system. For example, in many instances, the final or end product will simply be an updated or modified version of one that already exists. In fact, *Update O&SHAs* are typically performed routinely during a product's entire useful life cycle. Therefore, numerous procedures, operating instructions, and maintenance documents should already exist. Also, if it is truly a case of modification/update, then the end use of the product will most likely already be defined and the O&SHA can then be performed fairly early in the process.

In order to properly and accurately perform an O&SHA, the analyst should have access to detailed project descriptions and all appropriate design information. Operating sequence diagrams, functional diagrams, equipment panel layouts (if applicable), and other available drawings should be examined. Review of applicable regulatory codes and performance standards is also essential. The Preliminary Hazards List (if there is one), the PHA, and any subsequent SHAs are all excellent sources of documented information pertaining to the subject system which may prove as invaluable time-savers when performing an O&SHA. As previously mentioned, one primary objective of the O&SHA is to analyze maintenance procedures and operational documents for hazard control/elimination. These documents should therefore obviously be available for review. Finally, equipment operational performance data, facility peculiarities and other specifics, and information about the personnel (skills training, number of personnel involved, the size of the organization, etc.) that will operate/maintain the system must also be reviewed. Once the hazards associated with or generated by each of these performance and operational characteristics have been identified and evaluated, an O&SHA worksheet can be completed, as shown in Figure 8.1.

The final O&SHA report should include a description of the system or product, including any operating and maintenance organization/personnel and procedures. The main body of the report should provide a narrative concentrating on the key findings of the analysis with any recommended hazard control solutions. Usually, a calendar schedule showing anticipated dates of subsequent, future O&SHAs is included in the report along with the completed O&SHA worksheets (Stephenson 1991).

The properly executed O&SHA will address the total system interface and provide control measures for personnel interface hazards. It identifies all personnel and equipment within a determined hazard range of operations and evaluates caution and warning requirements. Special skills or additional training requirements are also typically identified in a completed O&SHA.

It cannot be overemphasized that the O&SHA is an extremely valuable analytical tool which concentrates on the human interface with the system, both from an operations and maintenance standpoint. Therefore, because of the critical requirement of ensuring that operations and maintenance personnel are protected from any hazards in their tasks, the importance of a properly performed O&SHA is of paramount concern in the system safety analysis process.

OPERATING AND SUPPORT HAZARD ANALYSIS

SYSTEM: _____

Operational Mode:_____   Performed By: _____

Page:_____   Date: _____

ITEM	PROCEDURE TASK	HAZARDOUS CONDITION	CAUSE	EFFECT	HAZARD LEVEL	ASSESSMENT	STATUS/ RECOMMENDATION

Figure 8.1   *Sample operating and support hazard analysis (O&SHA) worksheet.*

O&SHA EXAMPLE

To demonstrate the simple practicality of the *Operating and Support Hazard Analysis,* this example will evaluate the operator control console configuration for the vapor degreasing operation discussed in Chapter 6 (*Preliminary Hazard Analysis*). The reader will recall from Figures 6.6 and 6.7 that this operation required an operator to maintain a controlling position at a rigid console while manipulating panel controls to operate a crane, the vapor degreasing tank instrumentation, monitor operational timing, coolant flow and, at the same time, maintain visual contact with expensive parts during the performance of the degreasing task. All of these human-dependent tasks place a certain stress load on human performance and, indeed, some of the possible risk associated with operator error were initially identified on the PHL and PHA worksheets for this operation (Figures 6.10 and 6.11, respectively).

Scope and Purpose of the Example O&SHA

Using initial data identified during the development of the PHA, this O&SHA will only concentrate on the design and layout of the operator's control console. The analysis will focus on the actual placement of the console itself, as well as the location of the controlling dials, levers, indicator gauges, and switches on the main panel. If the possibility for hazard risk reduction exists due to simple panel/console redesign, the O&SHA will make the appropriate recommendations. This exercise is not intended to serve as a complete O&SHA for the vapor degreaser system. The purpose here is to provide a sufficient amount of information to illustrate the performance parameters of an O&SHA.

Risk Assessment

Figure 8.2 is a close-up view of the control console and main operator's panel as per current design specifications. The initial placement of the console itself, as well as the layout of the various dials, switches, gauges, and levers on the panel, as seen here, has been established based upon availability of panel space with minimal consideration of operator interface requirements.

Initially, the following risk potential exists if the current design were utilized:

- **Monitor Indicator Gauges**

  Hazard Condition 1: Gauge Similarity and Placement

  Cause: There are three indicator gauges which must be monitored by the operator of this console panel. Current design has each gauge nearly identical in their appearance, although their respective functions are entirely different. The three gauges, placed side-by-side, provide indication and operational status for solvent temperature, solvent liquid level in the tank, and coolant system pressure.

  Effect: Because of their location on the panel and their likeness to each other, there is a risk of operator misinterpretation or misreading of the respective gauges during task performance. This error could result in the following hazards:

  1. Solvent temperature inaccuracies (too high or too low). If the solvent temperature is too low, improper or inadequate liquid-to-gas conversion will result. This condition would have an adverse impact on the efficiency of the degreasing operation. If the temperature is too high, the solvent off-gassing rate would accelerate and either create an excessive amount of vapor in the tank (waste of solvent) or cause solvent vapors to escape the tank creating a hazard to personnel.

  2. Solvent liquid level in the degreasing tank could be inadequate (too high or too low). If the liquid level were too low, there is a potential that insufficient levels of vapor will be generated to accomplish the required amount of degreasing. If the solvent liquid level were too high, there is the possibility that the liquid would contact the metal parts and render them useless.

  3. Insufficient coolant pressure could result in a lack of coolant flow through the system, which in turn would effect the vaporizing action in the tank. Vapors could escape into the room and threaten personnel health by displacing the oxygen level in the room. If the pressure were too high, there is a slight risk of line rupture, although this is highly improbable due to excessive system design safety factors.

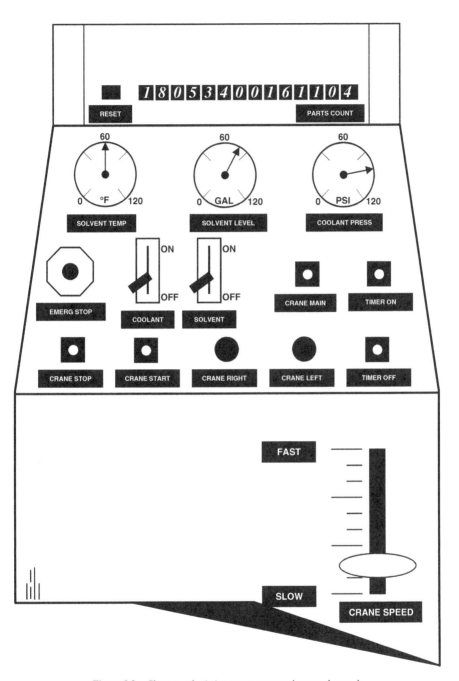

Figure 8.2   *Close-up of existing crane operator's control console.*

Risk Assessment 1: 1B

A hazard level of 1B has been assigned since, based upon the current panel design, it is quite *probable* that such an operator error could occur and the effects (threat to human health, equipment/parts damage, etc.) are considered *catastrophic*.

Recommendation 1: Relocate and redesign gauges, as indicated in Figure 8.3. Position all indicator devices for the solvent system on one side of the panel, and all coolant controls on the opposite side.

- **Activation of Critical Systems**
  Hazard Condition 2: Switch Location and Design

    Cause: The control switches for coolant flow and solvent flow are located side-by-side and are identical in appearance. As with the gauge assessment above, the result of this design scheme would be an increased potential for operator error in system activation.

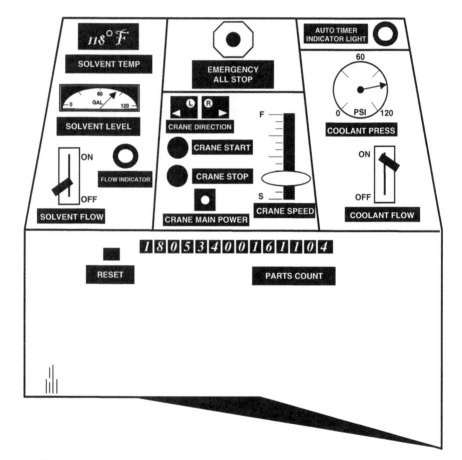

Figure 8.3 *Close-up of modified crane operator's control console after the O&SHA analysis.*

Effect: Inadvertent activation or deactivation of either system, at the wrong time during the operational flow, would adverse hazardous effects as follows:

1. Deactivation of the coolant flow instead of solvent flow would result in a steady loss of coolant in the system. This would allow Freon vapor to escape the tank and threaten human health.

2. Deactivation of the solvent flow instead of coolant flow would allow an excess amount of solvent liquid into the tank and subsequent equipment or parts damage is likely.

Risk Assessment 2: 1A

Due to the close proximity of one switch to the other, coupled with their identical construction and markings and the criticality in proper system operation, the potential for occurrence of human error in this instance is regarded as *frequent*. Due to the nature of the potential results (injury/death, property damage, and/or loss), this hazard risk has been categorized as *catastrophic*.

Recommendation 2: Relocate critical system control switches, as indicated in Figure 8.3. Position each switch on panel in close proximity to the system they are required to control (i.e., solvent and coolant).

- **Crane Operator Controls**
  Hazard Condition 3: Control Placement and Layout

  Cause: The crane control dials, buttons, and levers are literally scattered about the operator panel in the current design configuration. There appears to be no thought in the layout of these critical controls. The operator is subsequently required to maintain crane control while manipulating a variety of randomly placed control devices. For example, the speed control lever in the current design (Figure 8.2) is not even located on the main panel. Also, "crane left" and "crane right" controls are identical and placed opposite to and right next to each other (i.e., "control left" is on the right and "control right" is on the left).

  Effect: An inexperienced operator will have great difficulty remaining cognizant of all critical crane control functions. An experienced operator would also have some problem operating this panel if the current design layout is used, especially under stressful or emergency conditions. The resultant effect could be a loss of parts into the tank, inadvertent activation of similar-looking control dials and knobs, and subsequent misdirection of the crane load.

Risk Assessment 3: 2B

Based upon the current design layout, there is a high potential for *frequent* crane control errors. Since the resultant effect of this hazard would cause damage or loss of parts, as well as processing schedule delays, the hazard category of *critical* has been assigned.

OPERATING AND SUPPORT HAZARD ANALYSIS

SYSTEM: Vapor Degreaser –Parts Prep.

Operational Mode: Normal Configuration   Performed By: John Doe

Page: 1 of 1   Date: 29 April 2013

ITEM	PROCEDURE TASK	HAZARDOUS CONDITION	CAUSE	EFFECT	HAZARD LEVEL	ASSESSMENT	STATUS/ RECOMMENDATION
1	MAIN CONSOLE OPERATION	Gauge similarity and Placement	1. Solven Temp. Errort	1. Injury; Solvent Loss	1B	Operator Error Probable; Results Catastrophic	Relocate and Redesign Controls
			2. Solvent Liquid Level Error	2. Injury; Loss of Parts			
			3. Coolant Pressure Error	3. Injury; Line Rupture			
2	MAIN CONSOLE OPERATION	Switch Location and Redesign	1. Coolant System Activation Error	1. Injury; Parts Damage	1A	Operator Error Frequent; Results Catastrophic	Relocate and Redesign Controls
			2. Solvent System. Activation Error	2. Parts Damage or Loss			
3	CRANE OPERATOR CONTROLS	Control Placement and Layout	1. Operator Error During Crane Operations	1. Misdirection of Crane	2B	Operator Error Frequent; Results Critical	Relocate and Redesign Controls

Figure 8.4 *Crane operator's control console O&SHA worksheet.*

Recommendation 3: Redesign and relocate crane control devices to one central location on the panel. Eliminate the use of similar dials or buttons as much as design will permit. Figure 8.3 shows a suggested new layout. In addition, different color switches can be used for critical functions such as "crane stop" (suggest a red switch) and "crane start" (suggest a green switch).

Each of the above listed concerns have been recorded on the O&SHA worksheet as shown in Figure 8.4. From here, the analyst can prepare the O&SHA report which will detail each of these findings in much the same manner as provided in this example.

Of course there is potential for additional operator error as a result of this design. These include, but are not limited to

Condition: Less than obvious location of the Emergency Power Shut-Off control.

Recommendation: Relocate and isolate the control to center of the panel.

Condition: Current procedure requires operator to manually activate a timer once the parts are in the tank and deactivate it when the parts are retrieved.

Recommendation: Install automatic timer that will activate a flashing light indicator to alert the operator when the timing sequence has expired.

Condition: Operator cannot see parts basket over the top of the control panel because of the location of the parts counter mechanism on top of the console cabinet.

Recommendation: Relocate parts counter mechanism into the face of the console cabinet (Figure 8.3). Since this is only a passive indication device, its location on the panel itself is not critical.

Condition: Current labeling of control devices and indicators is difficult to read.

Recommendation: Use larger, more legible labels for control panel devices.

SUMMARY

The *Operating and Support Hazard Analysis*, or O&SHA, as discussed here is an integral part of the system safety analysis process. It is especially useful when evaluating operations or tasks that rely heavily on human performance. The O&SHA indicates that system design must consider the *ergonomic* or human-task interface element to ensure the identification and elimination or control of some types of hazard risk. The O&SHA is normally developed during the *design phase* of the project life cycle, using inputs from a variety of sources including the preliminary hazard analysis. However, the O&SHA can also be performed during the *operations phase*, especially after a system modification has occurred. If the subject system is particularly maintenance-dependent, the O&SHA can also identify potential hazard risk resulting from the human–machine interface that must occur during system servicing.

Where human task performance is concerned the importance of hazard risk reduction cannot be overemphasized. The O&SHA is an excellent tool to ensure the proper and adequate identification of such risk and to provide recommendations for risk reduction/control.

9

Energy Trace and Barrier Analysis

INTRODUCTION

The Energy–Barrier Concept

The *Energy Trace and Barrier Analysis (ETBA)* developed as a component of the *Management Oversight and Risk Tree (MORT)* program as a means of providing adequate analysis of accident cause. As discussed in greater detail later in Chapter 13, the MORT program suggests that accidents are usually multifactorial in nature. In the MORT program, an *incident* is defined as an unwanted flow of energy resulting from inadequate barriers or having a failure without consequence. An *accident* is further defined as an unwanted flow of energy or an environmental condition that results in adverse consequences. Hence, an accident can occur because of a lack of adequate barriers and/or controls upon an unwanted energy transfer associated with the incident (DOE SSDC-4 1983; Stephenson 1991). The accident is usually preceded by initiating sequences of planning or operational errors that eventually result in a failure to adequately adjust to changes in human factors and/or environmental factors. Failure to adequately adjust to these unplanned changes leads directly to unsafe conditions and unsafe acts that arise out of the risk associated with the subject activity. These unsafe conditions and unsafe acts, in turn, provoke the flow of unwanted energy. When proper barriers are not in place or are not adequate to control this energy flow, the resulting unwanted consequences (or accident) will undoubtedly occur. The

Basic Guide to System Safety, Third Edition. Jeffrey W. Vincoli.
© 2014 John Wiley & Sons, Inc. Published 2014 by John Wiley & Sons, Inc.

ETBA has been designed as an investigative tool with which to focus specifically upon four primary areas of concern.

1. Energy source(s) within a given system;
2. The adequacy of any barriers or controls within the energy path;
3. The human factors interface; and
4. The eventual *target(s)* of unwanted/uncontrolled energy flow (Note: *Targets* may be people or objects).

Uses of the ETBA

The ETBA is an analytical technique which can be of great assistance in the preparation of the preliminary hazard list (PHL). It can also be quite useful in the development of a Preliminary Hazard Analysis (PHA), Subsystem Hazard Analysis (SSHA), or the more general System Hazard Analysis (SHA). The ETBA can also be used, depending on the specific system under consideration, in the development of the Operating and Support Hazard Analysis (O&SHA), and, of course, during the MORT process from which the ETBA evolved.

In order to utilize the ETBA in the performance of the above listed system safety analyses, certain essential data are required for evaluation. For example, if the ETBA is to be performed on a specific manufacturing facility, then the analysis should begin with an examination of completed facility drawings. If the ETBA is concerned with a specific project, or a newly designed piece of manufacturing equipment, the project plans and schematics must be evaluated. It should be noted that the level of detail required is dependent upon the analysis itself. Development of a PHL will not require extensive detail and evaluation. Whereas an ETBA in support of an SSHA will meticulously analyze the project to the component level and detailed drawings will, therefore, be required.

Performing the ETBA

The ETBA begins with the identification of the types of energy that will be associated with the project, program, or equipment. Types of energy which can result in unwanted transfers include mechanical (rotating gear assemblies), electrical (energized systems or subsystems), potential (spring-loaded devices), kinetic (swinging armatures or a moving crane hook), natural (wind, temperature, solar radiation), radiation (ionizing, nonionizing, radiofrequency, electromagnetic), even biological (pathogens and viruses have energy), and so on. The next step in the ETBA process is to locate the exact source of the energy as it initially enters into the system or process. From here, the analyst will literally *trace* or plot the energy flow throughout the entire system. Next, wherever it may be possible for the flow of energy to contact a target element (i.e., people or objects), adequate barriers or controls must be in place. The ETBA will then identify all locations within the system where barriers are required, evaluate the adequacy of existing barriers, determine the risk associated with each unwanted flow of energy and assign a Risk Assessment Code (RAC) to that condition, and recommend barriers where none currently exist. Once completed, the ETBA will facilitate any subsequent evaluation of risk levels that may remain in the system after recommended barriers are installed (i.e., *residual risk*).

The ETBA Worksheet

Once the various types of energy affecting the system have been identified, the ETBA worksheet should be completed. Figure 9.1 shows a sample ETBA worksheet. The information recorded on the completed ETBA worksheet can then be used to perform subsequent analyses (PHL, PHA, etc.) along with their related reports. In some cases, depending upon the level of detail desired, the ETBA itself may provide an adequate amount of information to be included in the final PHA. In fact, since hazardous events can usually be associated with some type of energy transfer and, since accident causal factors typically involve the absence of controls or the failure of existing barriers and,

ENERGY TRACE AND BARRIER ANALYSIS

PROGRAM: _____   DATE: _____

ENGINEER: _____   PAGE: _____

DRAWING NUMBER	ENERGY AMOUNT & TYPE	BARRIERS (CONTROLS)	POTENTIAL TARGETS	RAC NO.	ANALYSIS OF BARRIER EFFECTIVENESS	RECOMMENDED ACTIONS	CONTROLLED RAC	APPLICABLE STANDARDS

Figure 9.1   *Sample energy trace and barrier analysis (ETBA) worksheet.*

since the resulting effect(s) of a failure on system operations are often determined in relation to the target of the unwanted energy flow, the information on the ETBA worksheet can actually be transferred quite easily to the PHA worksheet discussed in Chapter 6.

ETBA EXAMPLE

As stated earlier, the ETBA has great utility in determining the specific breakdowns in energy barriers during an accident/incident investigation. The ETBA is also quite useful in the analysis of new or existing systems to examine the adequacy of energy barriers currently in place.

The following example will evaluate an existing oxygen supply system installed at a fictitious hospital (*Memorial General Hospital*).

System Description

The oxygen system at Memorial General provides a high purity oxygen supply for use by individual critical care patients in their rooms, as well as during surgical procedures. The system consists of the following major components, as indicated in Figure 9.2:

- 3000 psi gaseous oxygen supply tank, located in a fenced-in protected area, outdoors, behind the hospital.
- Isolation valves, located throughout the system as needed.
- Main Distribution Service Panel, located in the hospital equipment room; it functions to regulate the source supply pressure from 3000 psi to 250 psi. Regulators, relief valves, and pressure indicating gauges are part of the Panel.
- Seam-welded conduit tubing, one inch in diameter, is used throughout the system to transfer oxygen supply from the Main Distribution Service Panel to the various demand locations.
- Several In-Room Service Panels are part of this system. Each patient room and other needed locations are equipped with an In-Room Service Panel. The Panel consists of an isolation valve, regulator hand valve (to reduce supply pressure from 250 psi to a usable pressure of 20–30 psi), a pressure indicator gauge, a pressure relief valve, a spring-loaded check valve, and a capped interface port (for maintenance operations only).

The ETBA

The ETBA begins with an identification of the types of energy involved in this case. For the purpose of this example, only the two primary sources shall be examined here. Obviously, the very first concern is the presence of an extremely *high pressure* in a confined system. The gas in this example (oxygen) is *highly combustible* and, thus,

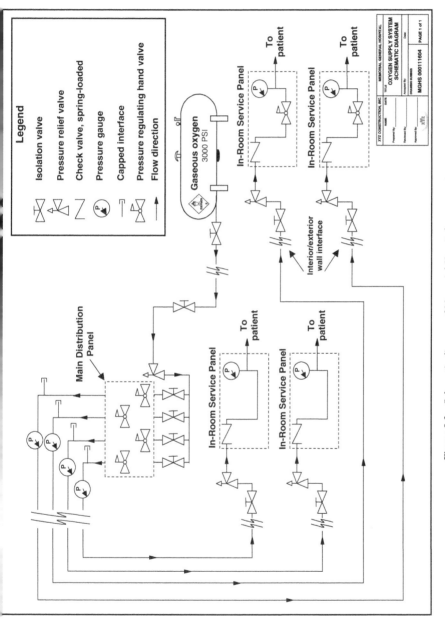

Figure 9.2 Schematic diagram of Memorial Hospital oxygen system.

ENERGY TRACE AND BARRIER ANALYSIS

PROGRAM: Oxygen Supply System   DATE: 07-15-2013

ENGINEER: Jane Doe   PAGE: 1 of 1

DRAWING NUMBER	ENERGY AMOUNT & TYPE	BARRIERS (CONTROLS)	POTENTIAL TARGETS	RAC NO.	ANALYSIS OF BARRIER EFFECTIVENESS	RECOMMENDED ACTIONS	CONTROLLED RAC	APPLICABLE STANDARDS
MGHS CMV-051834	High Pressure Gas	Isolation valves	Personnel	1C	Barriers adequate to prevent exposure	Ensure proper procedures are established and followed	1E	ASME OSHA 29 CFR 1910.134
		Fenced-off oxygen tank storage area	Personnel; Property	1B	Barriers adequate to prevent exposure	Increase security around hospital grounds	1E	
		Shatter-proof gauges and safety-related components	Personnel; Property	1D	Barriers adequate to prevent exposure	Ensure system design is per code	3E	ASME
		Distance from personnel and insulation inside service panels	Personnel; Property	2D	Barriers adequate to prevent exposure to high noise levels		3E	
MGHS CMV-111604	Oxygen-rich atmosphere	Field-welded tubing assemblies; scheduled inspections	Personnel; Property	1C	Barriers adequate to prevent exposure	Dye-penetrant inspection should be performed initially and every five years	2E	ASME

Figure 9.3   *Completed energy trace and barrier analysis worksheet for oxygen system.*

presents a second energy source of concern. Figure 9.3 reflects a partially completed ETBA worksheet for the following example analysis:

- **High Pressure Gas**
  Analysis of barriers/controls and potential targets:
  1. **Condition**: 1) The 3000 psi source oxygen supply is transferred into the hospital equipment room via a 2-inch supply line.
     Potential Targets: 1) Personnel contact with high pressure gas during maintenance and/or loading operations.
     Risk Assessment: 1) The result of this potential risk of hazard is considered *catastrophic* with an *occasional* opportunity for occurrence *(RAC 1C)*.
     Barriers/Controls: 1) There is currently an isolation valve in this line (on the exterior side) which enables complete energy system shut-off prior to energy flow into the facility.
     Controlled Risk Assessment: 1) Barriers are considered adequate with a *Controlled RAC of 1E* assigned.

2. **Condition**: 2) Oxygen source is a tank located outdoors.

Potential Targets: 2) Unauthorized entry into or tampering with tank supply area could threaten life and/or equipment.

Risk Assessment: 2) Results of such risk exposure would be *catastrophic* (possible loss of life) and the frequency *probable (RAC 1B)*.

Barriers/Controls: 2) The entire exterior oxygen storage area is fenced-off and locked shut to prevent unauthorized entry.

Controlled Risk Assessment: 2) Barriers considered adequate. Addition of security surveillance around the hospital grounds area would further reduce the probability of risk exposure. A *Controlled RAC of 1E* is assigned.

3. **Condition**: 3) Once in the equipment room, the oxygen supply enters the Main Distribution Panel. The Panel contains regulators, relief valves, and gauges to reduce the inlet pressure from 3000 psi to 250 psi.

Potential Targets: 3) Gauge or valve component failure could endanger operator and/or damage surrounding equipment.

Risk Assessment: 3) There is a *remote* possibility that such failures will occur with a potential hazard risk determined to be *catastrophic*, since serious injury or death could result from such exposure *(RAC 1D)*.

Barriers/Controls: 3) All gauges have been equipped with shatter-proof glass to prevent hazardous release of energy due to gauge failure. All valves have been rated and proof tested, in accordance with American Society of Mechanical Engineers (ASME) requirements, for use in this system.

Controlled Risk Assessment: 3) Design per applicable ASME codes have reduced the likelihood of such a risk to *improbable* and, if component failure were to occur, the shatter-proof gauges and rated valves would decrease the hazard category to *marginal*. A *Controlled RAC of 3E* is therefore assigned.

4. **Condition**: 4) In the event of pressure relief valve activation, potential dangerous noise levels may result in close proximity to personnel.

Potential Targets: 4) Operating personnel and/or patients may be subjected to damaging noise levels due to release of high pressure gas.

Risk Assessment: 4) The potential for unplanned or inadvertent system relief is considered *remote*, based upon system design factors, and the risk of hazard exposure has been determined *critical* since human health may be effected *(RAC 2D)*.

Barriers/Controls: 4) All relief valves are either isolated from personnel by distance or are contained inside an insulated service panel.

Controlled Risk Assessment: 4) Barriers adequate, isolation by distance acceptable and insulated control panels further reduce risk of exposure to hazardous noise levels. A *Controlled RAC of 3E* is assigned because, even if a relief did occur, the distance and isolation would create a marginal, rather than critical risk.

- **Oxygen Rich Atmosphere**

  Analysis of barriers/controls and potential targets:

  1. **Condition:** 5) Pure oxygen is piped throughout a hospital facility, between walls and ceiling/floor interfaces, and into various patient rooms and other needed areas.

     Potential Targets: 5) Hospital staff personnel and/or patients' possible exposure to oxygen-rich conditions creating a potential for fire/explosion and personnel injury/death as well as equipment damage/loss in the event of a system leak or failure.

     Risk Assessment: 5) The frequency of such an exposure, without controls, is considered occasional and the results, catastrophic (RAC 1C).

     Barriers/Controls: 5) All tubing and/or equipment connections have been field welded to ensure an adequate seal. System is visually inspected on an annual basis and hydrostatically leak tested every 5 years.

  Controlled Risk Assessment: 5) Barriers/controls considered adequate. The addition of a dye-penetrant analysis of all field joints would further reduce the possibility of hazard risk exposure. Enforced "No Smoking" policies around oxygen panels by the hospital staff and the patients will also decrease the risk potential. A *Controlled RAC of 2E* is assigned.

There are, of course, many other possible energy sources which would, under actual analysis, be evaluated for barrier adequacy. These may include *electrical* equipment located in close proximity of the oxygen system and relative location of *heat- or spark-producing devices* to the oxygen system.

SUMMARY

The *ETBA* is an effective system safety technique that can be used to evaluate the *adequacy* of existing or planned energy flow barriers with regard to hazard risk exposures. Typically, a well-developed ETBA is helpful in the development of a *PHA* for the entire system or project. The ETBA is most useful in *tracing* the flow of energy through a system to determine the cause factors that may have contributed to an accident or loss. However, there is also utility in the evaluation of existing controls to determine their value in preventing an unwanted flow of energy.

The ETBA is one of the fundamental tools of system safety analysis and, when used, cannot only document the adequacy of hazard barriers and controls, but it can also identify those energy flow areas within a system that may have been overlooked as potential risk hazards during the *concept* or *design phase* of the project.

10

Failure Mode and Effect Analysis

INTRODUCTION

The *Failure Mode and Effect Analysis (FMEA)* is one of the more familiar of the system safety analysis techniques in use. It has remarkable utility in its capacity to determine the reliability of a given system. The FMEA will specifically evaluate a system or subsystem to identify possible *failures* of each individual component in that system and, of greater importance to the overall system safety effort, it attempts to forecast the *effects* of any such failure(s). Because of the FMEAs ability to examine systems at the *component level*, potential single-point failures can be more readily identified and evaluated (Stephenson 1991). Also, although the FMEA should be performed as early in the product life cycle design phase as possible (refer Figure 3.4) based upon availability of accurate data, the system safety analyst can also use this tool, as necessary, throughout the life of the product or system to identify additional failure elements as the system matures.

Types of FMEAs

There are basically two types of Failure Mode and Effect Analyses. They are distinguished more by the *target* of the analysis than the actual analysis itself. In fact, the steps required in the performance of each are very similar, only the items being analyzed differ. Perhaps the fundamental difference between the two is in their *approach*. The first type, often referred to as the *functional* FMEA, utilizes the *deductive reasoning approach* (i.e., it begins by assuming a failure and focuses on the modes which

Basic Guide to System Safety, Third Edition. Jeffrey W. Vincoli.
© 2014 John Wiley & Sons, Inc. Published 2014 by John Wiley & Sons, Inc.

could cause that failure to occur). The second type, the *hardware* or *detailed* FMEA, uses the *inductive approach* by recognizing common failure modes and examining the effects of those failures on the entire system, its subsystems, or just one subsystem, depending upon the established scope of the FMEA (Stephenson 1991).

The functional FMEA targets any subsystems that may exist within an entire system. The functional FMEA will evaluate each subsystem and attempt to identify the effect of any failures in these subsystems. The analyst not only looks for the possible effects of subsystem failures on the system as a whole, but also examines the effect of such failures on other subsystems within the system. Although functional FMEAs are not as common as the hardware FMEA, their basic utility should not be dismissed. When a complex system (such as a nuclear reactor, an airliner, an overhead bridge crane, or a robotic milling machine) consists of numerous secondary subsystems, each with their own set of supporting subsystems, the functional FMEA should be performed to ensure proper system safety evaluation at every level.

The second and more common hardware FMEA examines actual system assemblies, subassemblies, individual components, and other related system hardware. This analysis should also be performed at the earliest possible phase in the product or system life cycle. Just as subsystems can fail with potentially disastrous effects, so can the individual hardware and components that make up those subsystems. As with the functional FMEA, the hardware FMEA evaluates the reliability of the system design. It attempts to identify single-point failures, as well as all other potential failures, within a system that could possibly result in failure of that system. Because the FMEA can accurately identify critical failure items within a system, it can also be useful in the development of the *preliminary hazard analysis* and the *operating and support hazard analysis* (Stephenson 1991). It should be noted that FMEA use in the development of the O&SHA might be somewhat limited, depending on the system, because the FMEA does not typically consider the ergonomic element. Other possible disadvantages of the FMEA include its purposeful omission of multiple failure analysis within a system, as well as its failure to evaluate any operational interface. Also, in order to properly quantify the results, an FMEA requires consideration and evaluation of any known component failure rates and/or other similar data. These data often prove difficult to locate, obtain, and verify (Stephenson 1991).

Performing an FMEA

To properly execute a Failure Mode and Effect Analysis, certain detailed data must be made available to the analyst. These data typically include, but certainly are not limited to, the following fundamental information for each system, subsystem, and their components (TAI 1989):

- Design drawings
- System schematics
- Functional diagrams
- Previous analytical data (if available)

- System descriptions
- Lessons learned data
- Manufacturer's component data/specifications
- Preliminary hazard list (if available)
- Preliminary hazard analysis
- Other system analyses previously performed

After the required information has been collected, the specific nature of the FMEA must be established. A firmly defined scope of the FMEA will assist the analyst in determining direction and ensure the FMEA remains in focus with these established objectives.

Once the scope of the analysis has been established, the FMEA can begin by examining the effects of specific failures in the system or subsystem. As these failures are identified, they are recorded on the Failure Mode & Effect Analysis Worksheet (Figure 10.1) for evaluation. The completed FMEA will then be very useful in the performance of other system safety analyses such as an SHA or the SSHA.

The FMEA Report

Upon completion of the FMEA worksheets, the analysis data are transferred into report format which should include, as a minimum, the following information:

Introductory Information: The analyst should provide basic information in this section of the report which describes the purpose and scope of the FMEA along with any limitations imposed on the analysis as a result (i.e., items not specifically within the scope of the analysis). The scope will also identify the type of FMEA (i.e., *functional* or *hardware*). Also included in the introduction section is an explanation of the methodology used to perform the analysis such as, but not limited to drawing reviews, examination of previous analyses (if applicable), evaluation of lessons learned, use of Preliminary Hazard List and/or Preliminary Hazard Analysis, and so on. Finally, any preestablished "ground rules" that may have been agreed upon should be provided here. Such ground rules typically limit or further narrow the scope of the FMEA, or just a portion of it, and should therefore be explained in the introductory pages of the report.

Definitions Section: Typically, an FMEA report will contain phrases or words that are not generally associated with the everyday practice of the industrial safety professional. It is therefore important to provide definitions and explanations of terms and phrases that will be utilized in the FMEA.

System Description: The FMEA report should contain a significant amount of descriptive information pertaining to the system or subsystem(s) being evaluated. The detail of this description is obviously dependent upon the available information. However, if the project or system is well into the design phase,

FAILURE MODE AND EFFECT ANALYSIS							
PROGRAM: _____			SYSTEM: _____			DATE: _____	
			COMPONENT: _____				
ENGINEER: _____			FACILITY: _____			PAGE: _____	
PART OR DRAWING	PART NAME	PART FUNCTION	FAILURE MODE AND CAUSE	FAILURE EFFECT ON SYSTEM OR COMPONENT	EFFECT ON JOB OR PERSONNEL	CRITICALITY LEVEL

Figure 10.1   *Sample failure mode and effect analysis (FMEA) worksheet.*

detailed information should be obtainable. This section of the FMEA report will explain the intended function or functions of the subject system under analysis. System components and required interfaces between components should be discussed in specific detail and to a level equal to that which will be required to understand the results of the FMEA. Caution should be exercised to remain within the established scope of the FMEA. The analyst need not provide too much or excessive descriptive information, particularly when it is not necessary to meet the objectives of the FMEA. This is especially true when the FMEA is only concerned with the evaluation of just one subsystem or even one component within a larger more complex system. If the FMEA is to be performed on an existing system, the description section should also include a detailed history of its use and performance to date, along with any noted safety-related concerns which may have been reported or documented in the past.

Criticality Assessment: This section of the FMEA report will detail the level of system, subsystem, or component criticality (refer Chapter 2, Table 2.3). This criticality assessment is usually based on some predetermined criteria that have been agreed upon by management. When evaluating a system during an FMEA, criticality is an expression of concern over the possible effects of a failure in that system. If such failures could result in adverse effects on personnel (e.g., death, serious injury, illness) or equally undesirable effects on equipment, components or the system itself (e.g., system loss, equipment damage), then the level of *criticality* will reflect this assessment.

Any and all *critical single failure points* (CSFPs) that were identified during the FMEA should also be provided in this section. The specific failure mode and its effect(s) should be listed and discussed here. The discussion should detail any acceptance or rejection rational to justify the recommended actions, which are provided later in the report.

Documentation List: To complete an FMEA properly, the analyst is usually required to review and/or reference many separate documents related to either the specific system being analyzed or the system safety task in general. Document numbers should be listed as well as any drawings or system specifications and schematics. Regulatory standards, if applicable, should also be referenced. Operating procedures, lessons learned documentation, vendor documentation, manufacturer's information documents/drawings, and so on, are all potential candidates for review and, if used in the development of the FMEA, should be listed in this section.

Data Section: All supporting data used to develop the FMEA, as well as that which can assist in the presentation of the final analysis, should be included in the final FMEA report. These data can include photographs of the system, subsystem, or individual component(s) being analyzed; layout drawings; electrical schematics; and, of course, the FMEA worksheets.

Critical Items List: The purpose of the FMEA is to identify and evaluate failure modes and the possible system effects of those failures. Since the potential for undesirable effects must be eliminated or controlled, the FMEA also provides recommended actions which must be taken to accomplish this goal. As part of this analysis process, the FMEA identifies any and all items within the system which, if a failure were to occur, would have a critical effect on the operation of that system. Therefore, to facilitate evaluation and analysis of these system effects, a critical items list is developed. The list provides detailed descriptive information on each item. It will explain its overall function within the system, as well as the function of any components that may make up that item. The failure mode determined as "critical" is then listed along with the potential effect(s) of such a failure. If an item on the critical items list is to be accepted as is, then acceptance rational must be provided. Such rational may include an explanation of any existing or planned design limitations that will prevent the failure during actual system operations, or the provision of excessive factors of safety that will render such failure(s) extremely improbable. Another area for evaluating acceptance is the history, or lack thereof, and any known failures of systems similar in nature and operation. Finally, the most important element of the entire report, the FMEA provides recommendations for management acceptance or rejection of the risk associated with any failure of any item on the critical items list.

FMEA EXAMPLE

To further understand the use of the FMEA, the following example of an overhead bridge crane will be evaluated. It is again noted that this analysis, as with other sample examples of analyses discussed in this text, is only provided in an effort to demonstrate the utility of a specific system safety analysis tool. It is therefore superficial in presentation and will only examine a select few of the many possible failure modes associated with the described overhead bridge crane system.

This example will develop a *hardware* FMEA for a proposed system that is well into the design phase of the product life cycle. For informational purposes, it is given that a Preliminary Hazard Analysis (PHA) was previously performed during the early stages of the design phase of this system. The information from the PHA will be used to assist in the development of the hardware FMEA. It should also be noted that the nature of an FMEA requires evaluation of subsystems, subassemblies, and/or components. For this reason, more detailed and specific descriptive information is provided here than that which has been supplied for previous examples discussed in this text.

System Component/Subassembly Description

Subsystem: Hoist Assembly   The overhead bridge crane in this example consists of two hoists: a main hoist of 10 ton capacity and an auxiliary hoist of 1.5 ton capacity. The crane has powered trolley and bridge drives with control from plug-in type floor consoles. This crane will operate within a manufacturing facility that produces speed boats and small fishing craft. The crane is manufactured and will be installed by XYZ Cranes, Inc. Figure 10.2 shows a simplified layout of the crane system.

The 10 ton (main) hoist assembly contains two main hoist motor brakes, a 30 horse power main hoist AC motor, the main hoist magnetorque load brake, main hoist gear reduction assembly, and the wire rope drum assembly. The 1.5 ton (auxiliary) hoist assembly contains the same components as the 10 ton. However, the auxiliary hoist motor is rated at 18 horse power and a single auxiliary hoist brake is provided.

Component: Electric Hoist Motors and Controls

Component: Electric Hoist Motors and Controls   Both hoists have enclosed, non-ventilated, 220/440 volt, 3-phase, 60 hertz, 1200 rpm motors. The control used on hoist motions is referred to as the "static stepless magnetorque control" on the manufacturer's drawings. With this control, drive motor torque is controlled by means

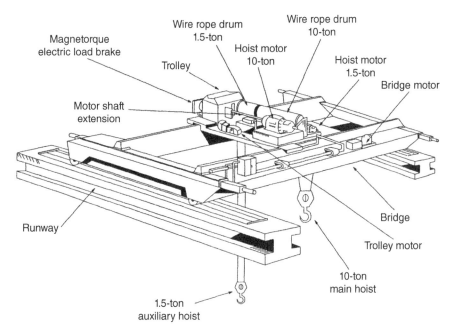

Figure 10.2 *The combination 10-ton/1.5-ton overhead bridge crane system.*

of fixed resistors and saturable reactors in the motor secondary circuit. To obtain very low speeds with overhauling or light loads, an electric load brake (magnetorque) is coupled to the motor shaft. The operator selects the direction and speed of the motion by moving the induction master handle from the "off" position toward the hoist (up) or lower (down) direction. There are two magnetic amplifiers, one responds only when the motor speed is above that called for by the induction master (operator controlled), the other when the motor speed is below that called for by the induction master. Normally, in the neutral or "off" position, the magnetorque load brake is fully activated. During crane movement, whenever the motion is slowed or stopped by moving the induction master handle toward "off," the magnetorque load brake provides braking torque which slows down the drive before the motor brake sets. This reduces wear of the motor brake shoes and wheels. After the motion has stopped, the "off" position bias circuit reduces the magnetorque load brake activation to minimize heating of the brake mechanisms.

Component: Magnetorque Electric Load Brake   The primary function of the magnetorque load brake is to incorporate an electrical brake to preload the motor and provide speed control without the use of a mechanical load brake system. The brake transmits torque by means of electromagnetic fields, there being no mechanical connection between stationary and rotating components. The stationary component is a doughnut-shaped coil, rigidly mounted and centered about the motor shaft extension. The rotor, or rotating component, is mounted on the motor shaft extension and rotates at motor speed. The breaking torque is accomplished through magnetic lines of force between the stationary field member and the rotor.

An aluminum fan, bolted to the rotor, draws air through the unit to remove heat from the rotor. In case of electric motor brake failure, the load will overhaul the hoisting unit. Then, the magnetorque brake will exert a braking torque to slowly lower the overhauling load to the floor, thus preventing a free falling load. With 90% of rated capacity load, the lowering speed will be limited to approximately 40% of maximum rated hook speed. The 10 ton hoist's rated hook speed is 0 to 7 feet per minute. The 1.5 ton hoist hook speed is rated at 0 to 20 feet per minute.

Component: Motor Brake Assembly   The motor brakes are spring closed, electrically released magnetic boxes. These brakes are function devices, utilizing brake shoes operating on the motor shaft extension. The magnetorque load brake slows motion to a very low speed before the motor brake sets and continues to apply braking torque until the motion stops, thus eliminating excessive wear and heating.

Component: Hoist Gear Reduction Assembly and Wire Rope Drum   The 10 ton main hoist gear case provides a 167 to 1 gear ratio through four gear reductions. The motor pinion shaft transfers power from the hoist motor to the main pinion shaft through a series of two intermediate gear and pinion assemblies. The main pinion shaft transfers power to the drum gear through the drum gear pinion resulting in the wire rope drum hoisting motion. The 1.5 ton auxiliary hoist gear reduction assembly functions in the same manner as the 10 ton assembly, however, the main pinion shaft drives the wire rope drum directly, resulting in one less reduction and a gear ratio of 125 to 1. Both drum assemblies are protected by gear driven upper and lower limit switches.

Subsystem: Motor-Driven Power Wheel   The motor-driven power wheel is mounted on the bridge assembly and functions as a "care taker" of the electric cable. It provides automatic operation as it rolls up, or releases, the cable as required by crane travel. It is powered by a drip proof, brake-equipped AC motor protected by a limit switch.

Subsystem: Trolley Drive Assembly   The trolley drive assembly moves the hoist assembly laterally across the 14 feet span between the runway rails. This is accomplished by applying torque through the trolley drive gear case to the trolley wheels. The assembly consists of a five horse power, 220/440 volt, three-phase, 60 Hertz enclosed motor, driving through the trolley drive gear reducer assembly. The motor is reversible and therefore able to drive the trolley in either direction. An electrically released, spring set motor brake provides frictional torque to brake and hold the trolley. Contact limit switches are provided for trolley travel in either direction.

Subsystem: Bridge Drive Assembly   The bridge drive assembly moves the main hoist (10 ton hoist) assembly along the runway rails. This is accomplished by applying torque to two of the bridge wheels. As with the trolley drive assembly, this assembly consists of a five horse power, 220/440 volt, three-phase, 60 Hertz motor with a motor brake, driving through a gear reducer assembly. This motor is also reversible and able to drive the bridge in either direction. The brake is spring set and electrically released.

When the motor is de-energized, the brake becomes set by allowing the brake spring to engage the brake and hold the bridge assembly. Contact limit switches are provided for bridge travel in either direction.

Subsystem: Control Station   Hoist control is accomplished from a plug-in type floor console which contains the following components:

- Main and auxiliary hoist controls
- Bridge and trolley controls
- Stop-start push button controls
- Selector (key) switch (master cut-off control)
- Run-inch push button controls
- Warning bell push buttons

Subsystem: 1.5 Ton Crane Micro-Drive System   During micro-drive operations (slow speed), an electric clutch is used to engage the micro motor and associated gear train. This allows smooth, minute movements of the hoist during critical operations.

Passive Components   Since an FMEA is concerned primarily with the identification of critical single-point failure items, concentration is usually specific to those subsystems and/or components that carry the greatest potential for failure. These items are generally those which move, are energized, cause other components to move or be energized, or, in some way would transfer its energy (kinetic or potential) in a given *failure mode* in such a way as to have undesirable consequences (i.e., *effects*). For this reason, the analyst may find it helpful or even necessary to list any and all passive components of a system, especially one as seemingly complex as an overhead bridge crane.

For the purpose of this example, the following components of the 10/1.5 ton overhead bridge crane are considered passive and will not be analyzed in the FMEA:

- Hooks
- Wire ropes*
- Load block
- Sheaves
- Suspension frame assembly
- Wire rope drums
- Drum gear
- Trolley and bridge shafts, couplings, and wheels (between the gear reducer assembly and the structural assembly)
- Structural components (rails, grinders, etc.)

__Note__: Although it is possible that a failure of a wire rope might impart or release some energy, depending upon the configuration at the time of the failure, it is highly unlikely due to required manufacturing design safety factors and will therefore be considered a passive component in this example.

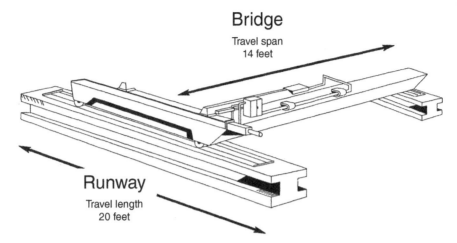

Figure 10.3   *Runway and bridge travel distances for the combination 10-ton/1.5-ton overhead bridge crane.*

System Operation

The overhead bridge crane system described above is used to transport precast fiber-glass speed boats and light fishing craft from their final assembly location, inside a manufacturing facility, onto a transport vehicle which is usually a flat-bed truck that has been backed into the facility. The operator must control the crane movement with the precision required to lift, transport, and lower the boats into place on the transporter. The hoist trolley design configuration will allow a travel span of 14 feet laterally, which provides ample operating flexibility when positioning the hook over the load, as well as when lowering the load into position on the truck. The longitudinal runway travel length of 20 feet also provides the operator with discretionary maneuverability during crane operations (Figure 10.3). The main hoist is used for movement of the boats themselves while the smaller, auxiliary hoist is primarily used to handle individual boat components being installed in the boat (e.g., seats, windshields, dash control panels).

Failure Mode(s) and Effect(s)

At face value, based upon the information provided, the complexity of the above described system appears to offer numerous opportunities for critical single-point failures. Therefore, the analyst should begin the FMEA process by first attempting to identify any and all nonpassive components and/or subassemblies that, depending on the type or mode of failure, could possibly have an undesirable effect. Table 10.1 lists each identified subsystem and related components in the overhead bridge crane system used in this example. Once such a list has been developed, the analyst will find it much easier to evaluate each subsystem or component, its possible failure mode(s) and the resultant effect(s) or any failure. Also, it is typical that subsequent

TABLE 10.1 Overhead Bridge Crane Subassembly Components

Subassembly	Components
Hoist assemblies	Hoist motor brakes
	30 Horsepower hoist motor (main)
	18 Horsepower hoist motor (auxiliary)
	Magnetorque load brake
	Hoist gear reduction assembly
Electric hoist motors	Fixed resistors
	Saturable reactors
	Motor secondary circuits
	Induction master
	Magnetic amplifiers
	Motor shaft extension
Magnetorque electric load brake	Electric brake
	Rotor
	Aluminum fan
	Field assembly
	Spring mechanism
	Magnetic brakes
	Friction devices
	Brake shoes
Hoist gear reduction assembly	Main pinion shaft
	Intermediate gear and pinion assemblies

FMEAs may be performed during the operational phase of the crane life cycle (refer Chapter 3, Figure 3.4). These subsequent FMEAs may only analyze a single subassembly or even a single component of specific concern within that subassembly. This is especially true when a modification to the crane takes place and the system safety effort is interested in analyzing the effects resulting from a failure of only that changed component, and not the entire system. Therefore, having developed a list similar to that provided as Table 10.1 during the design phase, the performance of any subsequent FMEAs on single components or subsystems will be greatly facilitated.

Evaluation of Potential Subsystem or Component Failures

Of the many possible failures that could occur within this system, the following examples are provided and discussed to demonstrate the typical approach used in developing an FMEA. Based on these examples, the reader should be able to grasp the fundamental FMEA concepts and, with some practice, utilize this tool to evaluate virtually any simple or complex system used in their respective organizations.

The completed FMEA worksheet (Figure 10.4) is provided to demonstrate its use in evaluating the following examples of identified failure modes.

- **Main and Auxiliary Hoist**
  Failure Mode 1: Hoist Inoperative

  The possible causes of an inoperative hoist include a loss in the power supply, defective circuitry, and/or defective bearings. (Note: In the examination of

FAILURE MODE AND EFFECT ANALYSIS

PROGRAM: Boat Manufacturing   **SYSTEM:** Overhead Bridge Crane   **DATE:** 29 September 2012

COMPONENT: Trolley & Bridge

ENGINEER: Jane Doe   **FACILITY:** Main Manufacturing   **PAGE:** 1 of 1

PART OR DRAWING	PART NAME	PART FUNCTION	FAILURE MODE AND CAUSE	FAILURE EFFECT ON SYSTEM OR COMPONENT	EFFECT ON JOB OR PERSONNEL	CRITICALITY LEVEL
XYZ Crane Drawing 04291954-B	Main and Auxiliary Hoist Motors	Provides motive power for raising and lowering suspended load from hoist	INOPERATIVE: Loss of power; Defective circuitry; Defective bearings	Load cannot be raised or lowered. Brake will hold load stationary.	No effect, except delay in operations during repair	3
XYZ Crane Drawing 04291954-B	Main and Auxiliary Hoist Motors	Provides frictional torque for stopping and holding load when hoist motor is de-energized.	FAILS TO ENGAGE: Broken springs; Worn linings	Load holding torque of motor brake will be lost. Redundant motor brake with electric load brake and motor control will hold load.	No effect, except delay in operations during repair	3
XYZ Crane Drawing 04291954-B	Main and Auxiliary Hoist Motors	Provides frictional torque for stopping and holding load when hoist motor is de-energized.	FAILS TO DISENGAGE: Loss of electric power.	Load cannot be raised or lowered. Brake will hold load stationary.	No effect, except delay in operations during repair	3
XYZ Crane Schematic No. CV34	Main Hoist Gear Reduction Assembly	Transfers motor and braking torque and provides for mechanical advantage through gear reduction from the motor to the hoist drum	DISENGAGES: Structural failure of gears, pinions or keys.	Torque necessary for lifting or holding load would be lost. Load would drop.	Possible loss of life or serious injury; damage to equipment and/or facility	1

Figure 10.4   *The partially completed failure mode and effect analysis worksheet for the crane system.*

listed failure modes, it is important to consider the cause(s) of any such failures. Once all potential causes have been identified, recommended corrective actions will be much easier to develop.)

Failure Effect 1: Load cannot be raised or lowered. Work stoppage may occur. No potential effect on other components or subsystems identified.

Assessment, Failure Effect 1: Noncritical potential. Crane load brake design will prohibit loss of load from a suddenly de-energized system. No identifiable adverse effect on personnel resulting from this condition.

- **Main Hoist Motor Brake Assembly (two in system)**

Failure Mode 2: Fails to Engage

The hoist motor brake assembly provides frictional torque for stopping the crane as well as holding a load when hoist motor has been de-energized. Possible causes of such a failure include a broken spring component or worn/damaged linings in the assembly.

Failure Effect 2: Simultaneous disengagement of both motor brake assemblies would result in the loss of holding torque of the motor brake which would, in turn, result in the loss of the load. Possible personnel injury/death and severe equipment/property damage or loss.

Assessment, Failure Effect 2: Since simultaneous disengagement of both braking assemblies is highly improbable, the effect of this failure is assessed as noncritical due the redundancy in this system. However, if this failure were to occur with no redundant electric load brake and motor control to hold the load, then the load would be lost and this failure would therefore have to be assessed as critical.

Failure Mode 3: Fails to Disengage

The main hoist motor brake assembly could fail to disengage if there is a loss of power during hoisting operations.

Failure Effect 3: The effect of such a failure during hoisting operations is the same as that discussed above in Failure Effect 1. The operator would not be able to move the load and, hence, operational delays would occur.

Assessment, Failure Effect 3: This failure effect is also assessed as noncritical since there will be no direct risk of hazard to personnel or equipment.

- **Main Hoist Gear Reduction Assembly**

*Failure Mode 4: Disengagement**

The disengagement of the main hoist gear reduction assembly is possible if there is a structural failure of the gears, pinions, or keys.

Failure Effect 4: The sudden disengagement of the main hoist gear reduction assembly would result in the loss of the torque used for lifting and/or holding the load. The result would be a loss of the load (suspended boat could drop). Possible personnel injury/death and severe equipment/property damage or loss.

Assessment, Failure Effect 4: Because there is no redundancy in this system to prevent the loss of the load, this risk is assessed as critical. Possible loss of life or the infliction of severe injury, as well as equipment or property loss could result from a dropped load.

*Note: It is noted that the very same potential failure mode exists with the auxiliary hoist as well.

In the above described failure mode and effects, one failure mode has been assessed as critical. The analyst should now provide either acceptance or rejection rational so that management can evaluate all aspects of this failure mode and effect before making a decision. Remember, one of the primary underlying purposes of the system safety effort is to provide management with choices in their evaluation of system risk. In this example, the following acceptance rational is suggested:

a. An operational check of the hoist will be performed before each use to verify proper operation of all gears. Such preoperational function tests will exceed the minimum acceptable testing required by OSHA at 29 CFR §1910.179(k)(1)(i).

b. Preventative maintenance inspections will be conducted weekly to ensure proper gear, pinion, and key operations. This periodic inspection interval will exceed the minimum monthly inspection defined by OSHA at 29 CFR §1910.179(j)(3).

c. Probability of failure is considered remote based upon the following design considerations:

  • Shaft and gear design will be in accordance with standards established by the American Gear Manufacturers Association.

  • The minimum design safety factor for this gearing will be 5:1.

  • All gears will turn in a continuous oil bath which will dissipate friction heat and lubricate the gear bearing surfaces.

  • Hook over-travel will be controlled by an upper limit switch.

  • Disengagement would require structural failure. The most probable result of structural failure would be binding and/or noisy operations which will alert the operator and those in the general vicinity of a problem with the crane.

d. A load test of the hoist at 125% of rated load will be performed annually. This annual test will exceed the minimum requirements for only an initial load test as specified by OSHA at 29 CFR §1910.179(k)(2).

e. There is no significant documented failure history of this type. In fact, standard industry reliability handbooks establish a generic failure rate for gears at 1.67 failures per million hours of operation.

Finally, based upon the results of the FMEA, including the presentation of the risk acceptance rational, the system safety analyst in this example is in the position of providing an accurate, well considered and properly evaluated recommendation to management to *accept the risk* associated with the failure mode of "disengagement."

SUMMARY

The *failure mode and effect analysis* is an excellent tool for evaluating the potential effects of failures within a designated system at the subsystem, subassembly, or component level. In fact, because of its ability to examine failure modes and their effects on either the individual component level (specific) or the subsystem/system level (general), the FMEA offers some level of flexibility in system safety analysis.

The *functional* FMEA is used to evaluate failures in one or many subsystems that function within a larger system, while the *hardware* FMEA examines failures in the assemblies, subassemblies, and components within those subsystems. The FMEA, therefore, has great versatility in the system safety process. The analysis can either be specialized, without regard for other subsystems which are not within the scope of the analysis, or it can be generalized to encompass total subsystem or system effects of a given failure condition. However, because the FMEA does not consider the human factors element or multiple failure analyses within a system, other types of system safety analysis tools and techniques should also be utilized.

11

Fault or Functional Hazard Analysis

INTRODUCTION

The *Fault Hazard Analysis (FHA)*, also referred to as the *Functional Hazard Analysis*, method follows an inductive reasoning approach to problem solving in that the analysis concentrates primarily on the *specific* and moves toward the *general* (TAI 1989). The FHA is an expansion of the FMEA (Stephenson 1991). As demonstrated in the previous chapter, the FMEA is concerned with the critical examination and documentation of the possible ways in which a system component, circuit, or piece of hardware may fail and that failure's effect upon the performance of that element. The FHA takes this evaluation a step further by determining the effect of such failures upon the system, the subsystem, or personnel. In fact, when an FMEA has already been completed for a given system and information on the adverse safety effect of component or human failures is desired for that system, the safety engineer can often utilize the data from the FMEA as an input to the FHA.

Although it can be performed later in the product development cycle than the PHA, the maximum benefit from an FHA is obtained if it is properly performed in the early stages of system development. The minimum requirements of an FHA are as follows:

- *Must* consider all functions;
- *Must* consider all functional failure modes;
- *Must* consider all operational phases;

Basic Guide to System Safety, Third Edition. Jeffrey W. Vincoli.
© 2014 John Wiley & Sons, Inc. Published 2014 by John Wiley & Sons, Inc.

- *Must* consider all operational interfaces;
- *Must* derive the failure condition and classify its severity;
- *Must* be systematic and thorough.

The FHA may derive hazard control criteria or even performance criteria where none previously existed. It may also establish the exact applicability of mandated criteria (standards and regulations). However, because of its ability to determine actual applicability of specific criteria as well as verify that maximum allowable probabilities are correct, the FHA is often useful in the analysis of a fully operational system as well. Like the SHA and SSHA, the FHA will examine small components or events to determine potential impacts on safety and reliability of the system or subsystem. The FHA requires a detailed evaluation of the system or subsystem and examines

- component hazard modes;
- causes of those hazards;
- resulting effects on the system or subsystem.

More simply stated, the properly performed FHA will attempt to answer at least the following two questions with respect to system or subsystem components (Larson and Hann 1990):

1. How can a component (or set of components) fail?
2. What will be the effect of this failure on the system or subsystem?

A fully developed FHA can also be used to define inputs to vendor design requirements, identify depth and scope of other analyses, define system architecture, and identify installation requirements and constraints.

The FHA Process

The FHA process usually begins with the establishment of a list of system or subsystem functions. Hazards are then postulated based upon the failure and/or likelihood of failure of each function. Then, the overall probable effect of the hazard upon the system and those operating it (i.e., people) is derived. Once identified, this overall probable effect is known as the *failure condition.* The severity of the failure condition is assessed and a hazard severity classification is assigned to it. This severity class will determine the maximum allowable probability for each failure condition. In extremely critical systems or operations, such as an elevator braking device/system or material handling operations, for example, very low maximum allowable probabilities identified in an FHA will mandate the prohibition of single-point failures.

Another benefit of performing the FHA in the early stages of design is the identification of Fault Tree Analysis "top events" (the failure conditions). Once the top events are defined, an inductive fault tree can be developed for each failure condition or event associated with the system. Chapter 12 will discuss the Fault Tree Analysis technique.

It should also be noted that each component of any given system may have more than one possible failure mode. Each mode may or may not become an additional hazard to normal operating conditions. Therefore, the resultant effects (e.g., damage, malfunction) on the system or subsystem are usually expressed in terms of probability and severity.

In this simplified explanation of the FHA, it should be evident that much of the data obtained during the PHA and, if performed, the SHA and/or SSHA can also be used to assist in the development of the FHA. The primary difference lies in the detail of the FHA method and the fact that it examines *all* component failure modes and assesses the impact of such failures. The potential impact to normal, safe operations may be negligible to none or devastating to catastrophic. The FHA, then, investigates the hazards of all system or subsystem faults.

The FHA which has been conducted during the early stages of product or system development should be re-performed whenever significant information about the system is further developed. If the subject system is an operational system, it is recommended that the FHA be performed on a periodic bases (e.g., quarterly, annually), depending on the potential hazards of possible failures. A new FHA should also be conducted following any major engineering design changes or modifications to a given system.

Figure 11.1 shows an example of a Fault Hazard Analysis Worksheet which can be modified to a specific system or subsystem. An explanation of each column on the worksheet is also provided.

FHA EXAMPLE

An automobile brake system will be examined using the *fault hazard analysis* method to determine potential faults in the systems current design. Once again, this is only an example to demonstrate the utility of the FHA and it should not be construed as an all-encompassing analysis of an automobile brake system. In this example, two designs shall be presented. The first (Figure 11.2) shows the system as it is currently designed (Larson and Hann 1990). The FHA worksheet, presented as Figure 11.3, resulted in the redesign of the system, as shown in Figure 11.4 (Larson and Hann 1990).

System Description

The system in this example is a simple automobile brake system that allows both front and rear braking action when the driver applies pressure on the brake pedal. Figure 11.2 shows the current design. The single-cell master cylinder serves both the front and rear brake systems. As pressure is applied, the piston in the master cylinder compresses the brake fluid in the cylinder. This compression force is transmitted through the brake lines to each brake shoe (two front, two rear) causing activation of

SYSTEM FUNCTIONAL/FAULT HAZARD ANALYSIS (FHA)

SUBSYTEM/SYSTEM: _____

PROGRAM: _____   DATE: _____

ENGINEER: _____   PAGE: _____

SYSTEM COMPONENT NOMENCLATURE	HAZARD MODE DESCRIPTION	SYSTEM OPERATING MODE/PHASE	EFFECT OF HAZARD ON OTHER SYSTEM	FAILURE CONDITION (EFFECT ON OVERALL SYSTEM)	ENVIRONMENTAL FACTORS	HAZARD SEVERITY CLASS	OTHER FACTORS INFLUENCING HAZARD EFFECTS	HAZARD CONTROL APPROACH
A simple description of the component or its official nomenclature; include a part number if available.	A description of the type of component failure that will result in a hazard.	The operating phase or mode of the entire system being considered in the hazard evaluation.	Address the hazard directly; provide a description of the result of the component failure.	Describe the larger result of the hazard, which may reduce subsystem effectiveness or result in subsystem failure and/or total system failure.	If applicable, a description of the operating environment and its influence on the severity or probability of the hazard.	Address both the probability and the severity. (Refer to Figure 2.1 and Figure 2.2).	If applicable, describe any other factors which may have an influence (direct or indirect) on the probability orseverity of the hazard.	Provide a brief description of the proposed controls to resolve the hazard.

Figure 11.1   Sample system functional/fault hazard analysis (FHA) worksheet.

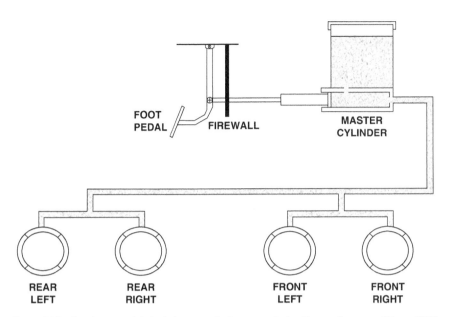

Figure 11.2 *Simple automobile brake/master cylinder system design (*Source: *Larson and Hann, 1990).*

the shoe and the resultant slowing of the wheel hub. This design offers unacceptable levels of risk, as will be shown during the following FHA.

The FHA Process

As stated earlier in this chapter, the FHA process begins with a listing of the system functions. In this example, the following functions can be expected from an automobile brake system (Larson and Hann 1990):

1. Put brakes on to stop car
   A. Linearly proportional to input force and travel (negligible timing delay);
   B. Withstand maximum pedal force without breaking;
   C. Capable of absorbing maximum brake energy;
   D. Not too much travel on pedal;
   E. Capable of maximum braking under maximum traction conditions;
   F. Left/right, front/rear brake force balanced;
   G. Work properly under all angular velocities;
2. Release brakes when pedal force is released
3. Capable of all the above in all environmental conditions
4. Self tightening.

SYSTEM FUNCTIONAL/FAULT HAZARD ANALYSIS (FHA)

SUBSYTEM/SYSTEM: _Automobile Braking System_

PROGRAM: System Design

ENGINEER: John Doe

DATE: _09-14-2013_

PAGE: _1 of 1_

SYSTEM COMPONENT NOMENCLATURE	HAZARD MODE DESCRIPTION	SYSTEM OPERATING MODE/PHASE	EFFECT OF HAZARD ON OTHER SYSTEM	FAILURE CONDITION (EFFECT ON OVERALL SYSTEM)	ENVIRONMENTAL FACTORS	HAZARD SEVERITY CLASS	OTHER FACTORS INFLUENCING HAZARD EFFECTS	HAZARD CONTROL APPROACH
Master Cylinder Proportional Braking Force	Loss of proportional braking force, causing all 4 wheels to turn freely.	1. Vehicle accelerating. 2. Vehicle accelerating. 3. Vehicle moving with engine off.	N/A	Vehicle has no brakes. Collision is likely, resulting in death or injury.	N/A	1A	N/A	System redesign to eliminate possible fault hazard.

Figure 11.3 _The completed system FHA for the automobile brake system (Source: Larson and Hann, 1990)._

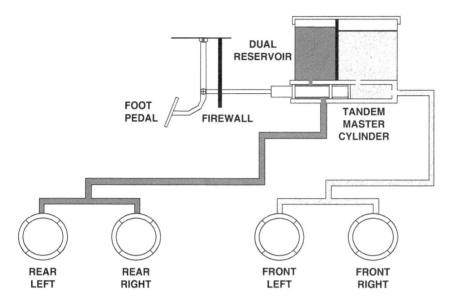

Figure 11.4 *Automobile brake/master cylinder system redesign following FHA* (Source: *Larson and Hann, 1990*).

There are many other expected functions of an automobile brake system that could be considered. However, for the purpose of this FHA, the above listed examples demonstrate the beginning of the FHA process for a given system.

During the next step in this process, the analyst will evaluate the current design to determine whether any faults exist that would prevent any or all of the expected functions from occurring under any operational condition. In this example, only Item 1A (proportional braking) shall be analyzed, and the results recorded on the FHA worksheet (Figure 11.3).

The FHA

Initially, the desired safety design criteria should be established. Generally, if the expected result of brake pedal compression is proportional braking, this is the basis of the design criteria for this item. A loss of all brakes is the event that is least desirable and, therefore, must be considered in the evaluation process.

Proportional Braking

Hazard Risk, Item 1: Loss of All Brakes As shown on Figure 11.3, column 2, the hazard mode is described as *loss of proportional braking force causing all four wheels to turn freely*. This hazard must be considered during vehicle acceleration, vehicle deceleration, and when the car is moving with the engine off (column 3). For this particular hazard, this hazard poses no discernible effects on other systems. However, the larger result of such a hazard (i.e., the *failure condition*) is listed as *vehicle has no brakes to stop; collision is likely, resulting in injury or death* in column 5 of Figure 11.3. There are no specific environmental factors listed here since total brake loss under any condition is an extremely dangerous situation. However, it is conceivable that a fault occurrence of this nature during adverse weather conditions (snow, rain, etc.) would greatly increase the loss potential and, thus, increase the level of risk.

Risk Assessment, Item 1: 1A   A hazard severity class is assigned a *catastrophic*, since the loss of human life is a potential result of this hazard risk. Because the current system design requires a single-cell master cylinder to serve both front and rear braking systems and a failure in the cylinder will permit a loss of all braking power, the probability for occurrence is listed as *frequent*.

Control Approach, Item 1   Redesign the system and incorporate a tandem cylinder which will serve front and rear brakes without compromising the safety of either if a fault should occur in the cylinder head. Figure 11.5 is a close-up view of the redesigned cylinder shown in Figure 11.4 which depicts the entire system, after the FHA inputs.

This example demonstrates the first evaluation of those functions which are expected to occur for an automobile braking system. The same analytical steps should be repeated for each item on the expected functions list until all fault hazards have been evaluated and controlled or eliminated. The FHA process ends when the last item on the list has been properly considered and/or resolved.

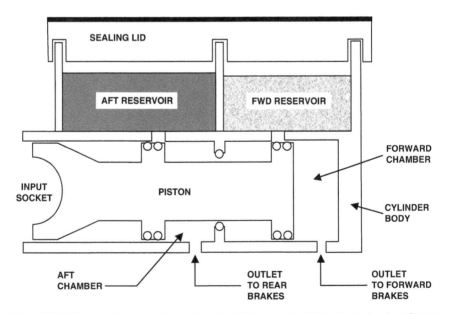

Figure 11.5   *Close-up view of the redesigned master cylinder, showing dual hydraulic chambers* (Source: *Larson and Hann, 1990*).

SUMMARY

It should be understood that the performance of an FHA is not always a requirement. Other analytical methods such as, but not limited to, the FMEA and the ETBA, if performed, should have already evaluated most, if not all, the same hazards which would be identified in the FHA. However, the FHA is a powerful tool in hazard identification and control and the benefit of its performance should not be overlooked. The FHA is an excellent system safety engineering method which can be used to ensure system operational integrity.

12

Fault Tree Analysis

INTRODUCTION

The *Fault Tree Analysis (FTA)* is considered one of the more useful analytical tools in the system safety process, especially when evaluating extremely complex or detailed systems. Because it utilizes the *deductive* method of logic (i.e., moves from the general to the specific), many system safety analysts find the FTA very useful in examining the possible conditions which may have lead to or influenced an undesirable or desirable event. As most occupational safety practitioners who have ever participated in an accident investigation know, undesirable events seldom occur as a result of just one initiating factor. For this reason, in the system safety process of fault tree analysis, the undesirable event is referred to as the *top event*. This is the *general* or known outcome of a possible series of events, the nature of which may or may not be known until investigated. As the analyst begins to identify the *specific* events which contributed to the top event, a fault tree can be constructed. By placing each contributing factor in its respective location on the tree, the investigator can accurately identify where any breakdowns in a system occurred, what relationship exists between events, and what interface occurred (or did not occur, as the case may be).

Although the FTA, by its very name, implies that it is primarily a tool for analyzing faults in a system or process, it is important to note that the FTA can also be used to evaluate the actions necessary to result in *desired* events, such as no accidents. By building a tree depicting all the events which must occur in order to realize that top event, the analyst can use the FTA as a method to construct the foundation of an

Basic Guide to System Safety, Third Edition. Jeffrey W. Vincoli.
© 2014 John Wiley & Sons, Inc. Published 2014 by John Wiley & Sons, Inc.

industrial safety accident prevention program. In this text, both uses of the FTA (i.e., evaluation of positive and negative top events) will be discussed through relatively simple and basic examples.

In the practice of system safety, the FTA is a very organized, meticulous, and versatile type of analysis. It is *organized* because it evaluates each event in consideration of that event's specific purpose, function, or place within a system or process. The FTA is *meticulous* because it attempts to describe the relationship of any and all events that may have acted upon a system to result in the top event. This method is also quite *versatile* in its ability to allow for the evaluation of hypothetical events, which the analyst may introduce into the tree to determine potential effects on the top event. Because of the flexibility offered by the FTA, it is most often used during the design phase of the product life cycle. The FTA can forecast potential failures in current design and identify areas where improvement is needed. It also has utility during the operational phase to determine the nature of real or potential desired or undesired events resulting from system operation. As is the case with most of the tools and techniques in the system safety process, the FTA will provide management with more alternatives when evaluating operational safety against operational cost.

Qualitative and Quantitative Reasoning

After all the causal events are listed on a fault tree, the FTA allows the analyst to evaluate each event separately or in combination with other events on the tree. This provides the user with a powerful tool capable of determining, through deduction, which event or set or events led to the top event. When more than one contributing event is identified, as is usually the case, their respective location on the fault tree is referred to as a *cut set*. Identification and qualification of one or multiple cut sets within a fault tree facilitates the evaluation process. Essentially, the cut set isolates specific events in the system and allows for a qualitative examination of the relationship between the set, as a whole, and its effect on the top event.

When the likelihood of an event is known and a probability value has been assigned, then analysis of these events on a fault tree will also yield quantitative results. As cut sets are identified, the probability of occurrence as a result of cut set interactions can be quantified and the associated risk can be more readily evaluated.

Constructing a Fault Tree

In order to properly construct a fault tree, the analyst must first possess extensive knowledge of the system or process under consideration. If such knowledge is lacking, then the process must include in-depth participation from the design community, as well as from other applicable organizational elements within the company (e.g., quality and reliability, operations engineering, facility operations). The analyst must obtain a clear understanding of the thought process behind the design of the system, as well as any operational criteria that effect system output.

An understanding of the operational working environment, as discussed in Chapter 2 (Figure 2.1), is also essential. Possible causal event factors may exist within the working environment and must therefore be considered in the FTA process.

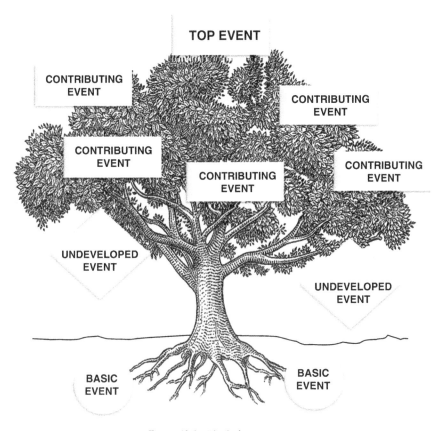

Figure 12.1 The fault tree concept.

The creation of a fault tree begins with the identification of the top event. This event can be as broad and general as *Total System Failure* or as narrow and specific as *Component X Malfunction*. This top event will be placed at the *top of the tree* and all subsequent events that lead to the main event will be placed as *branches* on the tree. Figure 12.1 illustrates the beginning of a simple fault tree, with the location of the top event, the placement of contributing events, and undeveloped events, down to the basic (or *root*) events. As the user moves from the top event downward, each level of the tree will materialize. In order to proceed from one level to the next, the analyst must continually ask the fundamental question: *What could cause this event to occur?* As causal events are identified, they are placed in position on the fault tree (Figure 12.1).

Fault Tree Symbols

Numerous symbols are used in the construction of a basic fault tree. These symbols, sometimes referred to as fundamental logic symbols, provide the analyst with a pictorial representation of the event and how it interacts with other events on the tree.

Figure 12.2 shows the basic symbols used during the FTA process. Once the reader has a general understanding of these symbols and their use, as described below, fault tree construction will be greatly facilitated.

The Rectangle   Used to identify the top or primary event, as well as secondary or contributing events (sometimes called main events). The rectangle shape as used on the fault tree indicates an event, or system state, which must be further analyzed on lower levels within the tree. This is the reason that the primary undesirable event is represented by a rectangle. Everything that appears under it is an attempt to further analyze its occurrence.

The Circle   Used to depict a basic event in the FTA process. It can be a primary fault event (i.e., the first in the process to have occurred) and, therefore will require no further development. Use of the circle symbol offers the analyst some flexibility. A causal chain could conceivably become quite extensive. Many times, the analyst will obtain sufficient casual information from analysis of higher level events in the chain. Therefore, in order not to waste valuable time and resources analyzing a single event to its lowest possible level, the analyst can label a particular event as basic, using the circle symbol indicating that no further development is required. For this reason, the symbols of the fault tree places the circle at the base of the tree (i.e., a "basic" event). The basic event is also often referred to as a "root" event or "root" cause, for obvious reasons.

The House   The house is used to identify a normal event which occurs during system operation. It is an event that either occurs or does not occur, such as turning a switch on or off. It should be noted that, if either the on state or the off state are possible during normal system operation, then the possible effect of both on the top event should be considered.

The Diamond   An event in the fault tree that is considered undeveloped is represented by a diamond. Use of this symbol identifies an event that the analyst has chosen not to develop further either because of the complexity of the event or because insufficient data are available to further analyze the event. Typically, it may also indicate an area of concern where further development should be considered at some point in the future.

The Oval   A conditional event or a conditional input into the logic tree that further defines the state of the system which must exist in order for the fault sequence to occur. It may place a restriction on event occurrence based on the occurrence of other events on the causal chain.

Logic Gates   A fault tree is basically a logic tree that shows the association between events on the tree. In general, there are two forms of logic gates which appear on a fault tree: The AND gate and the OR gate (Figure 12.2).

Use of the AND gate means that *all* contributing events connected to the main or primary events, through the gate, *must* occur in order for the main or top event to

SYMBOL	NAME	DESCRIPTION
	RECTANGLE	TOP EVENT; SECONDARY OR CONTRIBUTING EVENTS; A SYSTEM STATE REQUIRING MORE INVESTIGATION ON LOWER LEVELS
	CIRCLE	BASIC FAULT EVENT; NO FURTHER DEVELOPMENT REQUIRED
	HOUSE	NOT A FAULT EVENT; AN EVENT THAT IS EXPECTED TO OCCUR UNDER NORMAL OPERATION
	DIAMOND	UNDEVELOPED EVENT; ONE THAT, EITHER BY CHOICE OR NECESSITY, WILL NOT BE DEVELOPED FURTHER
	OVAL	AN EVENT THAT PLACES QUALIFIED CONDITIONS ON THE FAULT SEQUENCE
	AND GATE	DESCRIBES AN OPERATION WHERE ALL INPUT EVENTS MUST OCCUR FOR THE OPERATION TO OCCUR
	OR GATE	DESCRIBES AN OPERATION WHERE ONE OR MORE OF THE INPUT EVENTS CAN OCCUR IN ORDER FOR THE OUTPUT TO OCCUR
A1	TRANSFER GATE	USED TO SHOW LOGIC FLOW BETWEEN TWO PARTS OF THE FAULT TREE;TRANSFERS EVERYTHING UNDER THE EVENT IT IS ATTACHED TO; REFERENCE IS MADE BY AN ALPHANUMERIC CODE

Figure 12.2 *Standard Fault Tree Analysis (FTA) symbology.*

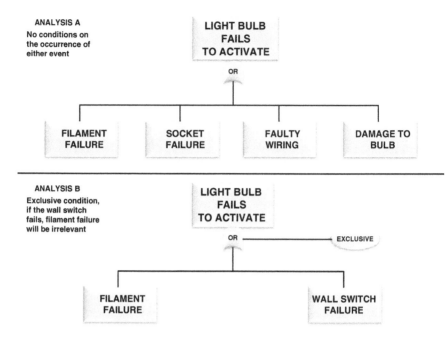

Figure 12.3   *The use of exclusive OR gates when proper conditions exist.*

occur. For example, when an AND gate is used, if only one or two of three listed events occur, then the main event will not occur.

The OR gate tells the analyst that if *either* event connected to a main event through an OR gate occurs, then the main event will also occur. This is an inclusive OR meaning that occurrence of any or all of the listed events will have the same result. There is an exception to this logic which must be understood. Sometimes, occurrence of one event connected to an OR gate, might *exclude* the possibility of other events occurring. For example, in order for a simple electric light bulb to function as intended, it has to be turned on. Events which could lead to failure of the light bulb to activate include filament failure, socket failure, damage to the bulb, faulty wiring, and so on. All of these events would be connected to the main event through an OR gate as shown in Figure 12.3, Analysis A. However, if the scope of the analysis is to consider only one of these possible failures, along with a failure of the light switching mechanism in the fault tree, the gate is now conditional and must be represented by an exclusive OR gate. The oval symbol, attached to the gate indicates the condition of the OR gate in this example (Figure 12.3, Analysis B). If more complex systems are evaluated, the exclusive condition placed upon this gate would require further explanation.

FTA Examples

As stated earlier, two types of fault trees will be constructed to demonstrate the use of this system safety analytical technique. The first, which will be referred to as a

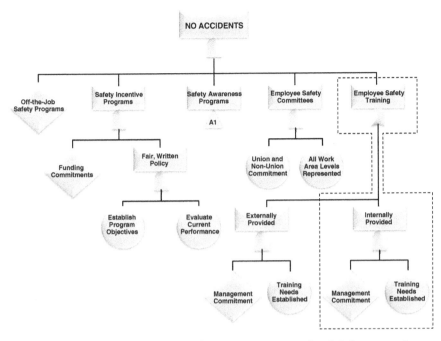

Figure 12.4 Sample fault tree analysis (FTA) showing structure, event and symbol placement, and cut-set identification.

positive fault tree analysis, will identify the events necessary to achieve a top *desired* event of no accidents. The second, or *negative* fault tree, will be constructed to show those events or conditions which will lead to a top *undesired* event of a fire in a manufacturing facility.

In the first example, the system safety engineer has been asked to assist in the industrial safety accident prevention program. By helping to identify those fault areas in the program where, if proper consideration is not provided, certain events could jeopardize the successful achievement of the primary objective (i.e., no accidents).

To construct this fault tree, the top event of *"no accidents"* is placed in the rectangle at the top of the tree. Refer to Figure 12.4 during this portion of the FTA example. The next step is to identify those areas or events that are necessary to achieve this top event. Basic industrial safety accident prevention program elements such as employee safety training, safety incentive programs, safety awareness programs, employee safety committees, and off-the-job safety programs are some of the many possible contributors to an accident-free work place. In Figure 12.4, the analyst has placed these sample events in rectangles under the top event. Because the analyst has determined that success in each of these areas is necessary in order to achieve the top event, an AND gate is used to connect them to the top event on the fault tree.

Working to the next level under each of these events, the analyst will identify each required supporting event. Where only one of many possible events are required in order to accomplish the main event, an OR gate is used in place of an AND gate (Figure 12.4). In some cases, events will not be developed further due to inadequate information, or a determination by the analyst that no further development is necessary. In this case, a triangle shape is used to characterize the event. At the bottom of this simple tree, some basic events have been identified and placed in a circle. The completed fault tree analysis for the top event of no accidents is somewhat self-explanatory. The FTA provides a road map that will lead to an accident-free work place. *Cut sets* leading to the top event can also be identified, as shown on Figure 12.4 by the dotted line. Of course further development of this tree will yield many more possible contributing and supporting events. However, for the purpose of demonstrating the construction of a *positive* FTA, this example should suffice.

In the second example, the *negative* fault tree with a top undesired event of fire has been developed as Figure 12.5. Here, the analyst shows the top event which, through the connection of an AND gate, cannot occur unless all of the three supporting conditions (ignition source, oxygen, and fuel) are present at the same time. This tree shows that each of these contributing events has additional supporting events under them which, if controlled, will not permit the main event to occur and, thus, the top event will not occur either. As in the previous example, this FTA also provides the user with a road map for controlling the outcome of the top event.

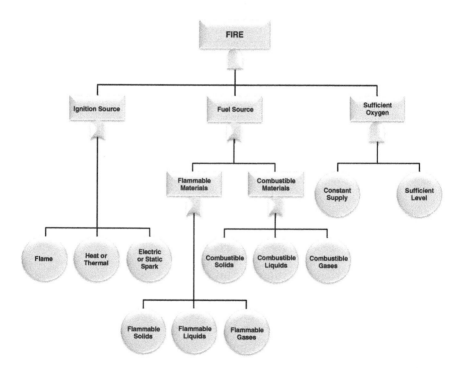

Figure 12.5 *Demonstrating the use of AND gates and OR gates in the building of a simple fault tree.*

Probability Values and the Fault Tree

As discussed in Chapter 5, known or deduced failure rates can assist in developing recommended control actions. In any given cut set, if a failure rate or a combination of failure rates are known to be quite low, then their potential effect on the top event can be numerically evaluated against the costs of controlling these risks.

A set of events attached to the main event through an OR gate are sometimes referred to as *mutually exclusive events* since the occurrence of the main event is not dependent upon the occurrence of all sub-events and, occurrence of each of the sub-events is not dependent upon the occurrence of each other. This means that, when events in a set are mutually exclusive, the probability of one or another of the events occurring is equal to the sum of the probabilities of the events occurring individually. This old but fundamental concept is known as the *addition rule* for probabilities (Spurr and Bonini 1973) and can be expressed as follows:

$$P(A \text{ or } B) = P(A) + P(B)$$

where P = probability; A = first possible event; B = second possible event.

On the fault tree, when probability rates are known, the analyst simply adds the probability values for the events under an OR gate and arrives at the expected probability for the occurrence of the main or top event. However, it is noted that in many systems, events are not mutually exclusive in that more than one can occur at the same time and result in the same outcome. In fact, the OR gate simply indicates that more than one or only one event must occur to effect the main event. When there is some probability that more than one event can occur at the same time to effect an outcome, it is known as *joint probability*. Figure 12.6 shows the overlapping effect which creates the joint probability theory. The event labeled AB in this diagram indicates that these events would actually be counted twice should the above formula be

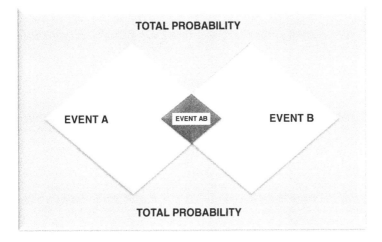

Figure 12.6   *The concept of joint probability of events. (*Source: *Spurr and Bonini 1973).*

TABLE 12.1   Injury Rates for a Sample of the Male and Female Population

	Men	Women	Total
Injury	17	3	20
Non-injury	53	27	80
Total	70	30	100

Safe behavior of 1000 men and women (percent of total)

used by accidentally labeling the event as mutually exclusive. In order to compensate for this potential, the formula can be modified slightly to allow for those events which are not mutually exclusive, as follows:

$$P(A \ or \ B) = P(A) + P(B) - P(A, B)$$

As an example: In evaluating the safe behavior of 1000 men and women, Table 12.1 shows the percent of the total that have either had occupational injuries or did not have occupational injuries. The contributing events (men having injuries and women having injuries) are not mutually exclusive since either or both may have suffered occupational injuries and, subsequently, effected the primary or top event (safe behavior). Figure 12.7 shows the fault tree for this extremely simple example. By applying the modified formula for non-mutually exclusive events, the probability of an injury event (I) involving a man (M) can be calculated as follows:

$$P(M \ or \ I) = P(M) + P(I) - P(M, I)$$
$$= 0.70 + 0.20 - 0.17$$
$$= 0.73$$

These data indicate that there is a 73% probability that an injury event will involve a male employee.

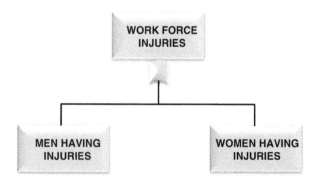

Figure 12.7   The top events of a fault tree analyzing safe work behavior.

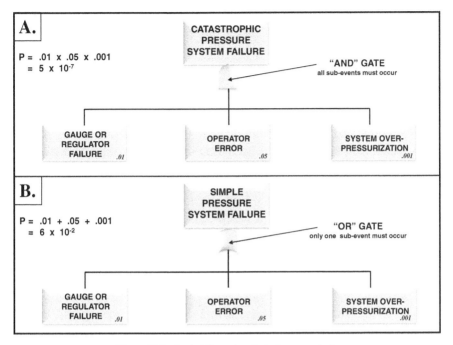

Figure 12.8 *Probability values in fault tree analysis.*

When a main or top event is the result of contributing events connected through an AND gate, the total probability of occurrence is, relatively speaking, somewhat less than if those events passed through an OR gate. Under an AND gate, each listed event must occur in order to realize the top event. Whereas, under an OR gate, only one of many possible events must occur for the top or main event to materialize. Therefore, to calculate the probable occurrence of an event supported by an AND gate, the underlying event probabilities are multiplied. For example, if a catastrophic failure of a pressure system (top event) is supported through an AND gate by three contributing events entitled *operator error, gauge/regulator failure,* and *system over-pressurization*, the simple fault tree could be drawn as shown in Figure 12.8A. Through manufacturer supplied data, gauge/regulator failure rates have been determined to be 0.01 based on 1000 hours of operation (10 failures in every thousand hours). Past accident/incident experience has determined the probability for operator error to be 0.05 (50 errors in 1000 hours of performance). Similar historical experience has shown that the probability of a system over-pressurization is remote at 0.001 (1 in a 1000 hours of operation). By placing these values under the appropriate labels on the tree, the analyst can already begin to evaluate the likelihood of realizing the top event (Figure 12.8A). Mathematically, the probability of experiencing a catastrophic failure of the pressure system is expressed as follows:

$$P(A) = P(B) \times P(C) \times P(D)$$

where: $P(A)$ = probability of catastrophic system failure; $P(B)$ = probability of gauge/regulator failure; $P(C)$ = probability of operator error; $P(D)$ = probability of system over-pressurization.

Therefore,

$$P(A) = 0.01 \times 0.05 \times 0.001$$
$$= 5 \times 10^{-7} (0.0000005)$$

Obviously, if all three sub-events must occur in order for pressure system failure to be classified as *catastrophic*, it appears that the probability of such a failure is extremely remote. However, if the analysis determined that an occurrence of any or all of the three events would result in *simple* system failure and that these events were mutually exclusive, then these events would be connected to the top event through an OR gate (as shown in Figure 12.8B) and the applicable probability values would be added, as follows:

$$P(A) = 0.01 + 0.05 + 0.001$$
$$= 6 \times 10^{-2} (0.061)$$

Hence, when occurrence of the top event is not dependent upon the occurrence of all sub-events, the probability of the top event is not so remote. In this example, the likelihood of simple system failure (as opposed to catastrophic system failure) has increased several orders of magnitude and would therefore require a more informed decision regarding hazard risk acceptance.

SUMMARY

The FTA is a technique which can be used to identify those events which can or must occur in order to realize a *desired* or *undesired* outcome. The technique uses a *deductive* approach to event analysis as it moves from the *general* to the *specific*. The FTA has great utility in its ability to distinguish between those events which *must* occur (represented by an AND gate) and those that simply *can* occur (represented by an OR gate) in order for the top event to occur. The information charted on a fault tree provides a *qualitative* analysis by demonstrating how specific events will effect an outcome. If probability data are known for these events, then the FTA can also provide *quantitative* information to further evaluate the likelihood of achieving the top event.

Once developed, the fault areas which are responsible for yielding an undesired (or desired) event can be evaluated on the *micro* rather than the *macro* level and this is one of the primary utilities of the fault tree analysis technique.

13

Management Oversight and Risk Tree

INTRODUCTION

The *Management Oversight and Risk Tree (MORT)* was originally conceived and developed in 1970 by W. G. (Bill) Johnson at the request of the Energy Research and Development Administration (today's Department of Energy, and the now defunct Atomic Energy Commission), Division of Safety, Standards, and Compliance. Prior to that time, there was no coordinated and formalized system safety program within the organization. The original proposal for the establishment of such a program made the following arguments (DOE SSDC-4 1983):

> .... Emerging concepts of system analysis, accident causation, human factors, error reduction, and measurement of safety performance strongly suggest the practicality of developing a higher order of control over hazards (than currently exists).

> .... The formulation of an ideal system appears to be a valuable precondition for knowing what information to seek after an accident and what aspects of performance to measure.

Johnson, who had recently retired as General Manager of the National Safety Council, advanced the idea that the application of controls and resources by the management of occupational safety and health programs could be categorized into five basic levels.

1. Less than minimal compliance with regulations and codes.
2. Minimal compliance with regulations and codes.

Basic Guide to System Safety, Third Edition. Jeffrey W. Vincoli.
© 2014 John Wiley & Sons, Inc. Published 2014 by John Wiley & Sons, Inc.

3. Application of manuals and standards.
4. Advanced safety programs exemplified by those currently found in leading companies.
5. An as-yet-nonexistent, higher level safety program synthesized by combining the "system safety" concepts pioneered by the military and aerospace industry with the best occupational safety practices and factoring in the newer concepts of behavioral, organizational, and analytical sciences.

Johnson was convinced that a sufficient amount of data already existed to suggest that progression from one level of safety program performance to the next higher (better) level might result in an order of magnitude reduction in the annual rate of disastrous accidents experienced by a specific enterprise. His extremely analytical approach to this study yielded the first generation MORT text in 1971. At that time, four key innovative features now basic to the MORT program were introduced.

1. An analytical "logic tree" or diagram, from which MORT derives its name.
2. Schematic representation of a dynamic, universal system safety model by using fault tree analysis methodology.
3. Methodology for analyzing a specific safety program, through a process of evaluating the adequacy of the implementation of individual safety program elements.
4. A collection of philosophical statements and general advice relative to the application of the MORT system safety concepts and listed criteria by which to make an assessment of the effectiveness of their application.

The MORT Analytical Chart

Throughout the early-to-mid 1970s, the "new" MORT concept was subjected to extensive studies and further development through practical application at test locations. The MORT program (or *programmatic MORT*) as it exists today, is viewed as a specialized management subsystem that focuses upon programmatic control of industrial safety hazards. The actual logic diagram (or *analytical MORT*), displays a structured set of over 1500 *basic events,* nearly 100 generic *problem areas,* and an unknown number of *judging criteria,* all interrelated as safety program elements and concepts comprising the ideal safety program model. Figure 13.1 shows the way in which the MORT concept (*programmatic MORT*) is schematically represented by a logic diagram (*analytical MORT*). The MORT chart has been increasingly used and accepted as a method for analyzing a specific accident or, alternatively, evaluating an existing safety program for accident/incident potential.

As a safety management program, MORT has been designed to prevent safety-related oversights, errors, and/or omissions by providing relatively simple decision points in an accident analysis or a safety program evaluation. The end results of programmatic MORT implementation are the identification, assessment, and referral of residual risks to the proper management levels for appropriate action. MORT will

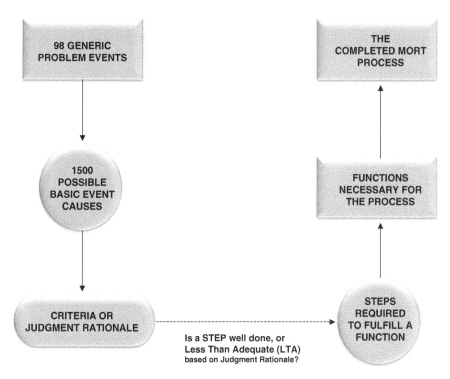

Figure 13.1 *The management oversight and risk tree (MORT) process.*

also serve to optimize the allocation of resources available to the safety program and to individual hazard control efforts.

MORT Use

Primarily, MORT has been designed for use as an investigative tool with which to focus upon the many factors contributing to an incident/accident. It accomplishes this by means of a meticulous trace of unwanted energy sources, along with consideration of the adequacy of the barriers provided. Hence, one can understand how the energy trace and barrier analysis (ETBA), as discussed in Chapter 9, grew from MORT as a completely separate analytical technique. MORT, however, goes much further than the ETBA. As the analysis proceeds, the MORT chart will identify any detected changes in the system (planned or unplanned). When change is detected, MORT recommends the performance of a detailed *change analysis* (DOE SSDC-4 1983).

When system changes are considered, their potential consequences must be evaluated in terms of risk acceptance. Here again, the appropriate management level must determine what is considered *acceptable risk*. Good business management will identify the need for proper control methods such as barriers to reduce levels of risk. MORT, then, is designed to investigate accidents and incidents and to evaluate safety programs for potential accident/incident situations. Two of the many basic MORT concepts are the analysis of change and the evaluation of the adequacy of energy barriers relative to persons or objects (i.e., *targets,* as discussed in Chapter 9) in the energy path.

THE MORT EVENT TREE

It is not possible within the limited scope of this text to provide fully detailed and comprehensive instruction on the use of the MORT event tree in either safety program evaluation or accident/incident investigation and analysis. Indeed, experts in this technique, such as Bill Johnson, have written numerous textbooks on just this system safety concept alone (see Appendix A). The event tree working model is detailed on a single chart 30'' × 24'' (without instructions). Reproduction of the entire event tree would require several pages alone which is not practical. Because of the complexity and overwhelming nature of the full MORT event tree, a *mini-MORT* chart has been developed to facilitate the analysis of relatively minor incidents, as well as to serve as a tool for instructor users in the MORT technique. The mini-MORT reduces the number of events to be analyzed and evaluated from 1500 to approximately 150 by eliminating the bottom tier of the chart and removing all transfers from the chart. Since the use of transfer gates to associate information from one chart to another is avoided in the mini-MORT, the overall chart tends to appear less complex and intimidating. Additionally, the seldom-used event symbols like the scroll and the stretched circle are replaced with the more common circles and rectangles (Stephenson 1991). However, with the mini-MORT, reproduction of the entire event tree would not be possible in a single-page text format. Other excellent alternatives to MORT and mini-MORT, such as Stephenson's *PET chart* (Project Evaluation Tree) also exist, but will not be discussed here.

In general, the remainder of this chapter will focus on the explanation of the various MORT event tree symbols and their use/meaning. Since the tree is an analytical model, the information presented in the previous chapter (Fault Tree Analysis) will be helpful and should be reviewed.

Symbols

The primary symbols used for most analytical trees have been used in the MORT event tree as well. These include

- the *rectangle* (primary or top event and secondary, contributory or main events),
- the *diamond* (undeveloped event),
- the *circle* (basic event),
- the AND gate,
- the OR gate,
- the *oval* (conditional or constraint symbol), and
- the *triangle* (transfer gate or symbol).

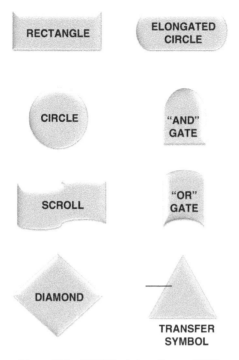

Figure 13.2   *MORT Symbology* (Source: *DOE*).

In addition to these, the MORT chart also uses a *rounded rectangle*, or *elongated circle*, to represent a satisfactory event (an event that may have contributed to an accident or incident but whose existence is essential for normal system operation). Also, instead of the *house* symbol common to fault tree analysis to represent those events that are considered normal and expected in a typical system, MORT uses a *scroll* symbol. Figure 13.2 shows the MORT symbols as described here.

MORT Analysis Example

In accident investigation, MORT analysis begins as soon as the accident/incident occurs. MORT moves from the known (accident event) to the unknown (causal factors) through an extremely complex, exacting, and quite meticulous *process of elimination*. The top event (e.g., injuries, damage, performance loss) is identified and assigned the appropriate position in the rectangle at the top of the event tree. Contributing events, or *blocks* of many possible contributing events, are placed under the top event in typical fault tree fashion. Figure 13.3 (DOE SSDC-4 1983) shows the top events of the MORT chart.

Construction and layout at this level depicts the main tree "branches" as Specific and Management (S/M) Oversights and Omissions on the left side of the tree, and Assumed Risks (R), on the right. MORT dictates that risk factors are defined as only

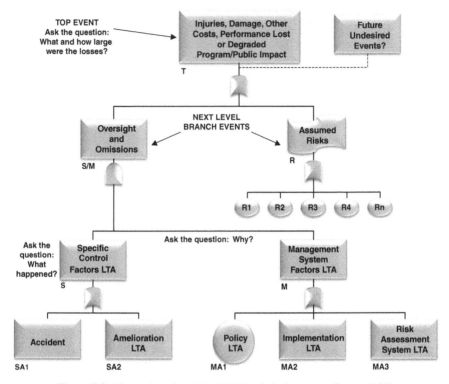

Figure 13.3 *The top branches of the MORT analytical event tree (Source: DOE).*

those risks that have been analyzed and accepted by the proper level of management. Nonanalyzed or unknown risks cannot be considered Assumed Risks. Because Assumed Risks have been accepted by management, MORT requires risk events in the Management Oversight and Omissions branch to be transferred to the Assumed Risks branch (since it may have been a management oversight to assume a given risk event).

The two next-level branches under Oversights and Omissions are labeled Specific Control Factors (S) on the left side or the branch and Management System Factors (M) on the right. On the MORT chart, Management factors are deliberately shown separate from the process that produced the specific adverse event for two reasons (DOE SSDC-4 1983) as follows:

1. Depiction of the existing management systems will suggest related background aspects of the specific accident that should be closely examined and
2. The specific event may, in turn, suggest certain aspects of the management system which may truly be less than adequate (LTA).

In order for MORT to be truly comprehensive, all related and seemingly unrelated events in the management system must be examined for possible contribution to the top event.

The MORT investigative process can be best understood through a detailed examination of each element on the MORT event tree diagram. The individual branches and the events assigned to them are somewhat self-explanatory in that each element of the branch asks a relatively simple question. Starting at the top of the diagram with the actual loss event (or the potential for a loss if MORT is being used to evaluate an existing safety program) and moving, in turn, through each of the three main branches, the analysis begins to identify, isolate, eliminate, and/or evaluate all possible contributory factors that may have influenced the top event. Detailed consideration of the Specific Control Factors branch is accomplished by reasoning backward in time, through several sequences of contributing factors. The analysis in this branch ends when the question posed by the basic event (circle) statement can be answered with a definitive "yes" or "no." Obviously, some factors (branches) will be more relevant than others. However, MORT deliberately has many planned redundancies throughout the diagram. A higher degree of hazard protection is attained when a hazard may be identified and connected at two or more places. It is better to ask the right question twice than to fail to ask it at all (DOE SSDC-4 1983).

MORT Color Coding

MORT investigation utilizes a color-coding system, as follows, to help identify those areas on the event tree where additional investigation or analysis is warranted.

Red: Event or factor *Less than Adequate* (LTA). Any event, factor, or block of events or factors, that, after thorough examination, has been determined to have LTA controls or barriers to prevent a transfer of hazardous energy is colored red on the MORT chart. Caution in the use of this color is warranted since any system elements labeled LTA must be well documented and a recommended course of action provided in the final accident investigation report to management.

Green: Event or factor *Adequate*. When investigative evaluation reveals that the barriers, controls, procedures, training, and/or any other factor effecting an event is considered to have been adequate (i.e., not likely to have contributed to the primary event in the block), the color green is used. Since green will also indicate that no further analysis will occur in this event block, the analyst should also cautiously use this color code.

Black: Event or factor *Not Applicable*. Depending upon the nature of the top event, there may be some areas of the comprehensive MORT event tree that simply do not apply to the particular investigation. In this case, the event and/or an entire event block is colored black or simply crossed-out. Although it may be obvious in many instances when a set of events is not applicable, the analyst must truly verify (not assume) that specific events are not contributory to the investigation. Inadvertent elimination of a possible contributing factor could jeopardize the overall analysis process.

Blue: Event or factor examined. *Insufficient* data available to fully evaluate the event or event block. Additional detailed analysis and investigation of these

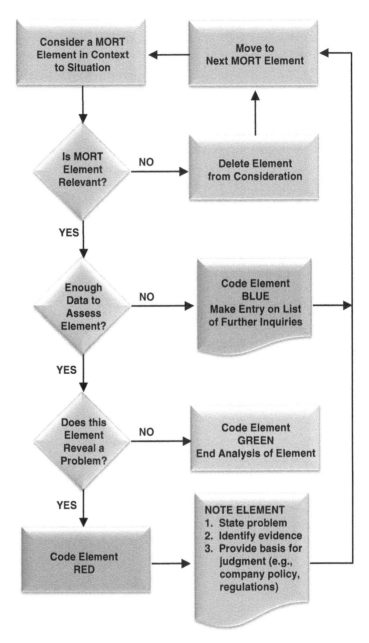

Figure 13.4   Sequence of work through a MORT Analysis Chart (Source: *DOE*).

areas may produce an adequate amount of data to determine the true value of these factors or events. Once this has occurred, these events will be re-coded with one of the other indicator colors. Ideally, the MORT process will exhaustively examine all blue areas to remove any question as to their value. However, in many cases, it is simply not possible to eliminate all blue events from a truly comprehensive investigation (sometimes, data are just not available).

The *MORT User's Manual* first published more than 30 years ago states that it is best to assume that all the boxes in the MORT diagram are colored blue, meaning that much more information is required before a judgment can be made (DOE SSDC-4, Revision 3 1992). This is where the questioning begins in the MORT process. Events can then be colored the appropriate indicator (red, green, or black) depending upon the answers obtained. The investigation is complete only when the entire chart has been evaluated through this question–answer process.

PROCEDURE FOR MORT ANALYSIS

Simply stated, the objective of a MORT analysis is to understand how specific "targets" were exposed to harm, damage, or unwanted change and to explain this in terms of risk management. To begin the analysis, an event is chosen from a previously performed ETBA (see Chapter 9) and is placed at the top of the MORT diagram. Then, the analyst considers each MORT element (e.g., oversight and omissions, assumed risks, specific control factors, management system factors) relative to the situation under analysis. By following the logic flow shown in Figure 13.4 in the analysis of that event, and for each subsequent event, the process will eventually lead to the root or basic causes of the chain of failures that resulted in the loss or damage.

SUMMARY

The MORT analytic diagram works best if it is used as a working paper. Pertinent facts about an accident or problem may be noted in margins at appropriate places. Informality is a key factor in the practice of MORT analysis. The diagram itself will ensure proper discipline in the performance of the analysis. MORT is a screening guide and a working tool, not the finished report. Writing the accident investigation report is a separate process. MORT is structured to facilitate analysis of a given accident event or the potential for its occurrence.

14

HAZOP and What-If Analyses

INTRODUCTION

The Hazard and Operability (HAZOP) study and the What-If analysis are analytical techniques that have their roots in the petrochemical industry. The HAZOP and the What-If analysis have become two of the most common *qualitative methods* used to conduct process hazard analyses in the petrochemical industry. In fact, up to 80% or more of a given company's process hazard analysis may consist of HAZOP studies and What-If analysis. The remaining 20% is comprised of checklists, fault tree analyses, failure mode and effects analyses, event trees, and so on. (Nolan 1994). Just as the foundation of system safety engineering had its roots in the emerging missile and aerospace industries during the 1940s and 1950s, this text has demonstrated that the methods and techniques associated with system safety have direct application in the practice of industrial safety compliance. Likewise, the principal concepts of the HAZOP study and the What-If analysis processes have definite application in general industry as well. An experienced review team can use these techniques to evaluate possible deviations from design, construction, modification, and operating intent to define and prevent potential unwanted consequences or cause desired outcomes to materialize. This chapter will present a brief introduction to the use of the HAZOP study and the What-If analysis in an effort to show how they are used in tandem to produce a comprehensive and highly detailed assessment of operational risk.

These analytical methodologies were first developed for the petrochemical industry with the intent to reduce the probability and/or consequences of a major incident

Basic Guide to System Safety, Third Edition. Jeffrey W. Vincoli.
© 2014 John Wiley & Sons, Inc. Published 2014 by John Wiley & Sons, Inc.

that would have serious or even devastating impacts to the employees, the public, property (both company-owned as well as off-site), and/or the environment. Extension of these concepts into a more broad application in industry in general could have far-reaching benefits and present certain advantages to an organization's overall accident/incident prevention programs.

BACKGROUND

The HAZOP study was first developed in the United Kingdom for use in its chemical industry during the 1960s. Imperial Chemical Industries, Ltd (ICI) is credited for developing this standardized approach to the analysis of process hazards associated with basic operating conditions of their facility. Then, using the HAZOP and What-If methodologies, changes to individual operating protocols were introduced (on paper) one at a time to allow the review team to evaluate the subsequent (albeit hypothetical) consequences. Over time, this analysis method evolved into a standard practice, first at ICI and then into the chemical industry in general. While it should be stated that HAZOP was not uniformly or consistently applied, the concepts still form the basis of the HAZOP approach that is in general use today. By inputting these hypothetical changes to the operating system, the potential consequences can be better understood and, if necessary, actions can be taken to preempt any possibility of realizing such consequences under real-world operating conditions.

Definitions

For the purpose of this text, the following definitions are provided for the HAZOP study and the What-If analysis:

HAZOP: A formal, systematic, logical, and structured investigative study for examining potential deviations of operations from design conditions that could create process-operating problems and hazards.

What-If Analysis: An informal but somewhat structured investigative method for introducing and evaluating hypothetical events, or series of events, associated with the operation of a given facility or process.

The successful use of both the HAZOP study and the What-If analysis is dependent upon the expertise and experience of the individuals that comprise the review teams. Essentially, both are really nothing more than exercises in communication. While each method can be conducted as separate analyses, the What-If analysis is almost always a primary component of a complete HAZOP study. Information is presented, discussed, analyzed, and recorded. Specific safety aspects and requirements are identified so that appropriate design considerations can be determined. The objective is accident prediction and the end result is accident prevention.

Objectives

There are basically four primary objectives of any HAZOP study or What-If analysis (Nolan 1994).

1. To identify the causes of all deviations of changes from the intended design function;
2. To determine all major hazards and operability problems associated with any identified deviations;
3. To decide whether action is required to control the hazard or operability problems;
4. To ensure that the actions decided upon are implemented and documented.

TEAM MEMBERS

Conducting a proper HAZOP study and any supporting What-If analysis is and should always be a team effort. The review team should consist of individuals that work in or are affected by the process under review. These can include supervisors, maintenance personnel, and line employees, as well as safety representatives and even consultants (when necessary) to ensure a well-rounded, experienced-based review team. In the end, the team should consist of at least the following "types" of members:

1. A team leader
2. A recorder (or scribe) to document all team activities, decisions, and reports
3. The "experts" (those who are experienced with the process, equipment, procedures, facility, etc., under review)

The ideal size of a review team should not exceed five or six members. This is because too many participants will make the review process particularly difficult to follow and document. The more members, the more the number of conversations, and the greater the potential for distraction and confusion.

REFERENCE DATA REQUIREMENTS

When performing either a HAZOP study or What-If analysis, it is essential that all necessary reference data be available for review and evaluation. Any and every type of data that pertains to the system or project under analysis should be examined to ensure that the review team is fully cognizant of all design criteria, functional intent, and operational requirements of the system. Once these elements are clearly established and understood, the consequences of any potential deviations (intended or otherwise) can be more fully explored and evaluated.

TABLE 14.1   Typical Sources of Reference Data to be Reviewed When Conducting a What-If Analysis and/or HAZOP Study

Examples—sources of reference data
• History of losses from the existing or similar facilities
• Ergonomic or human factor elements (color coding, accessibility, practical use, languages and instructions, etc.)
• Personnel staffing and headcount levels (distribution of personnel, levels of supervision, etc.)
• Evacuation routes and emergency response/action plans
• Design codes and applicable standards (ANSI, NFPA, OSHA, ASME, ASTM, ISO, HSE, etc.)
• Utilities specifications and reliability history/data (power, water, gas, etc.)
• Ambient environmental data (expected operating temperatures, prevailing weather conditions, seismic considerations, etc.)
• Special studies or calculations where available and applicable (vapor dispersions, blast overpressure, etc.)
• As-built drawings showing interface to exiting systems
• Instrumentation criteria and approach (local/remote control, hardwired/data highway, failure modes, analog/digital, emergency alarms, etc.)
• Facility and equipment electrical diagrams
• Data sheets for instruments and controls (i.e., vendor or manufacturer specifications)
• Full description of system design specifications and calculations
• Operating procedures including startup and commissioning as well as shutdown and disposal (if applicable)
• System maintenance history, including regular maintenance schedules and any known or previous emergency maintenance requirements
• If applicable, chemical and physical properties of all commodities involved in the system (material safety data sheets, UN specifications, etc.)
• Fire and explosion protection systems drawings and arrangements (fire and gas detection/alarm, protection—passive/aggressive, etc.)
• System design philosophy and requirements (e.g., process description)

Table 14.1 (Nolan 1994) provides an example listing of some of the more common reference sources typically identified for use in HAZOP and What-If analyses. The challenge is to identify and obtain *all* potential source data to ensure a comprehensive review.

THE CONCEPT OF "NODES"

In HAZOP studies and What-If analyses, the review team must first identify the areas or components of the system that will each be analyzed during the review process. In the chemical industry, these individual components are typically referred to as "nodes." There are three basic criteria for identifying the nodes to be reviewed (Nolan 1994).

1. Divide the facility into process systems and subsystems
2. Follow the process flow of the system under study
3. Isolate subsystems into major components which achieve a single objective (i.e., increase pressure, remove water, separate gases).

Using these criteria, individual nodes associated with the vapor degreaser system discussed in Chapter 6 (refer Figure 6.6) can be identified:

1. Compressor
2. Refrigerant storage tank
3. Cooling coil
4. Solvent storage tank
5. Overhead bridge crane
6. Electrical service

Once the individual nodes have been identified, the HAZOP study and What-If analysis can be initiated. As stated earlier, the information developed through the What-If analysis can then be used as input data for the more complete and detailed HAZOP study of the vapor degreasing system.

CONDUCTING THE WHAT-IF ANALYSIS

While both the HAZOP study and What-If analysis are generally organized and con-ducted in a similar fashion, the HAZOP study is more comprehensive and structured, while the What-If approach is generally broader and less formal. For these reasons, each method will be presented and discussed separately.

In the petrochemical industry, the What-If analysis has been found to be most useful in analyzing facilities and processes that are relatively simple in structure, operations, and overall complexity. Extending these "lessons learned" from years of application in the chemical industry to general use, it is suggested that the occupational safety practitioner interested in the utility of the What-If analysis attempt using this approach to analyze hypothetical consequences for relatively simple operations and procedures. As established previously in the text, the level of completed design can also dictate which types of analyses will be most effective. During the concept and design phases (see Figure 3.4, Chapter 3), only general information may be available. Hence, a detailed HAZOP could not really be performed. During these phases, when specific details of a project or process are not yet fully established, the focus should be on the prevention of accidents through analysis of specific events (i.e., those that have the potential to impact total system safety). The What-If analysis is one of many possible tools to accomplish this objective. It should also be noted that, depending on the specific scope of a particular analysis exercise, the What-If approach can be successfully used in all phases of the product life cycle.

What-If Analysis Steps

To initiate a What-If analysis, the review team must establish certain *assumptions* about the process or system under review. These assumptions help establish the boundaries of the overall review and ensure the analysis remains focused on providing value-added recommendations to project management. To illustrate the need and use of assumptions, the What-If analysis of the vapor degreaser, assumes the following:

1. Process equipment is compatible with materials used and is properly designed;
2. The facility is operated with an adequate and properly trained staff as intended by the design philosophy;
3. The failures of process equipment, instrumentation, and safety devices occur randomly;
4. The failure rates and demand rates of safety devices are considered to be low;
5. The safety of personnel and property and the preservation of the environment are top priorities for project management and employee labor representatives;
6. The facility is designed, operated, and maintained to good management and engineering standards.

With these assumptions in place, the review team will not waste valuable time pursuing issues such as the use of incompatible materials, operator training inefficiencies, high rates of equipment failure, or inappropriate environmental, safety and health (ES&H) management policies and directives.

The next step is to determine the types of failure events that are possible for the system under review. It is critical to differentiate between events that are considered likely to occur (credible) and those that are considered extremely unlikely (noncredible). Again, the purpose is to establish workable boundaries for the review and to ensure that the team stays focused on only those issues that will add value to the overall process. For the vapor degreaser operation, Table 14.2 provides examples

TABLE 14.2   Partial List of Credible Failure Events for the Vapor Degreaser Operation

Vapor degreaser operation	
Credible failure events	Examples
A single human error with or without established operating instructions	Incorrect sequencing of steps; improper positioning of controls; prolonged exposure of parts to degreasing vapors; materials moved from degreasing tank to quickly; solvent tank level not maintained
Two simultaneous human errors with or without established operating procedures	Incorrect sequencing of steps; improper positioning of controls; incorrect crane movement and/or function; unacceptable coolant and/or solvent temperatures
A single instrument or mechanical failure	Compressor failure; loss of coolant flow; loss of solvent flow; refrigerant tank pressure regulator failure; loss of cooling
A single human error, coupled with a single instrument or mechanical failure	Compressor failure; improper interpretation of console instruments; loss of coolant flow; loss of solvent flow; refrigerant tank pressure regulator failure; unacceptable coolant and/or solvent temperatures

TABLE 14.3   Partial List of Non-credible Failure Events for the Vapor Degreaser Operation

Vapor degreaser operation	
Non-credible failure events	Examples
Simultaneous failure of two essential independent instrument or mechanical systems	Malfunction or simultaneous failure of compressor and refrigerant system; loss of cooling and failure of flow level shut-off sensors in tank
Failure of both the crane automatic stop safety device and the crane magnetorque braking device	Loss of crane lowering control limits and inability to stop crane movement into tank
Simultaneous loss of main electric service power and backup electric generator power	Complete loss of power to compressor, crane, operator console, solvent heating system, refrigerant cooling system
Facility destruction due to uncontrollable external events	Earthquake, terrorist attack or civil unrest, flood, etc.

of credible failure events and Table 14.3 shows examples of those that would be considered non-credible and will, therefore, not be subjected to analysis.

The What-If Analysis Worksheet

As with numerous other analytical techniques presented and discussed in this text, a worksheet developed for the What-If analysis greatly facilitates the documentation and review of data points. Figure 14.1 provides an example of a What-If worksheet.

Figure 14.1   Sample What-If Analysis worksheet.

				WHAT-IF ANALYSIS WORKSHEET			

AREA: _____Vapor Degreaser_____   **TEAM MEMBERS:** _J. Doe; T. Smith; C. Hope; V. Jones; A. Ford_

DATE: _____29 September 2012_____   **PAGE:** _____1 of 1_____

ITEM	WHAT IF............	HAZARD	CONSEQUENCES	SAFEGUARDS	RECOMMENDATIONS	REMARKS
1	The wrong solvent is introduced into the system instead of Freon 113?	Contamination; undesired reaction	Equipment damage; product loss; waste disposal issues; production downtime	Vendor Certificate of Acceptance; sampling/testing before off-loading	Develop verification procedures prior to acceptance of product	None
2	Solvent unloading line fails?	Released of solvent liquid and vapor at high pressure	Environmental impact; personnel/ public (off-site) effects; public relations; product loss	Engineering controls; system design/layout; equipment; personal protective equipment; SOPs in use; controlled access; low pressure switch; remote switch; dikes/ containment	Investigate availability of alternative high pressure hoses and lines	Materials used now are not always compatible with chemical commodities
3	All lines, valves, or vessels are not identified as to contents?	None to catastrophic	None to catastrophic	Periodic inspections; common industry practice	None	None
4	Spare parts control procedure is not adequate?	None to catastrophic	None to catastrophic	Critical parts have been identified	Evaluate current purchasing practices for critical spares	None

Figure 14.2 *Partially completed What-If analysis worksheet for the vapor degreaser system.*

By placing specific questions in the "What-If ... " column, the analyst can evaluate the consequences of credible hypothetical situations and events.

Figure 14.2 shows the partially complete What-If analysis worksheet for the vapor degreaser facility. Note that many more "what-if" questions can (and should be) asked to completely exhaust all possible events that could effect the safety of overall system operation. Quite typically, a properly completed What-If analysis will address many dozens of individual items and encompass numerous worksheet pages before all credible events have been evaluated. The intent of Figure 14.2 is to demonstrate the use of the worksheet in a What-If analysis. For each expected use of the What-If analysis, it would be helpful if a What-If checklist is developed before-hand. Checklists will establish all possible questions that could be asked for a given system or process. This will facilitate the actual What-If analysis since all questions will be readily available for consideration. For example, in addition to the what-if questions appearing on the worksheet in Figure 14.2, additional questions to be asked may include the following:

What if facility electrical power is lost or interrupted?
What if the overhead crane magnetorque brake fails?

What if the refrigerant level is too low?

What if the system pressure is too low?

What if the system pressure is too high?

What if the solvent level is too high?

What if the solvent level is too low?

What if the facility heating, ventilation, and air conditioning (HVAC) system fails?

What if the main compressor fails?

What if the refrigerant tank relief valve malfunctions?

What if the sprinkler system activates inadvertently?

What if process flow diagrams or drawings are not current?

What if obsolete or unused equipment is installed?

What if the solvent tank fails?

What if the sanitary sewer system fails?

The questions should continue until all elements are addressed. If a more detailed and complete HAZOP study should be required, the information contained on the What-If analysis worksheets will help facilitate and streamline the pending HAZOP study.

CONDUCTING THE HAZOP STUDY

The more complex the system or process to be evaluated, the more essential is the need for a HAZOP study. The HAZOP study is conducted in much the same way as the What-If analysis, usually by the same review team. There are minor differences, however, in terminology and approach. In the HAZOP study, certain "guidewords" are normally used to aid the review team and help identify specific areas where deviations from design intent can occur. Guidewords can include pressure, flow, level, temperature, power, and so on. HAZOP also attempts to identify the severity of the outcome if such deviations from the norm occur as well as the probability or likelihood of occurrence. The Hazard Risk Matrix established and explained in Chapter 2 (Table 2.3) can be used for this purpose since it provides both severity and probability rankings for a given hazardous situation.

The HAZOP Worksheet

Taking the above stated differences between the What-If Analysis Worksheet and the HAZOP process, Figure 14.3 provides an example of a typical HAZOP Worksheet. Note the headings over each column. Some are identical to those found on the What-If worksheet and, therefore, transfer of the information from any previous What-If

HAZOP WORKSHEET								

AREA: _____   EQUIPMENT/PROCESS: _____   TEAM : _____

DATE: _____   PAGE: _____

ITEM	GUIDE WORD	DEVIATION	POSSIBLE CAUSES	POSSIBLE CONSEQUENCES	SAFEGUARDS	RAC	RECOMMENDED ACTION(S)	REMARKS

Figure 14.3   Sample HAZOP worksheet.

analyses would be relatively simple. Figure 14.4 shows the beginning of what should conceivably be an extremely detailed HAZOP study of the various systems and components (nodes) of the vapor degreaser system. A similar review of each node in the entire system will produce a comprehensive study of all possible deviations from design intentions. Once these are evaluated and understood, decisions can be made as to the safeguards that should or should not be employed.

THE ANALYSIS REPORT

Once all What-If analysis questions have been asked and answered along with all completed HAZOP studies of system components, a final report should be written to document all findings and recommendations. In the chemical industry (in the United States), this report is normally referred to as a Process Hazard Analysis. This report is required under both Occupational Safety and Health Administration (OSHA) and Environmental Protection Agency (EPA) regulations for facilities that handle or contain certain chemical commodities at certain defined quantity thresholds. However, when HAZOP studies and What-If analyses are used in general industry application, the documentation of the results can be included in a written report along with any other system safety analyses that may have been performed (as described

HAZOP WORKSHEET

AREA: Vapor Degreaser

EQUIPMENT/PROCESS: Freon 113 Transfer

TEAM : J. Doe; T. Smith; C. Hope; V. Jones; A. Ford

DATE: 29 September 2012

PAGE: 1 of 1

ITEM	GUIDE WORD	DEVIATION	POSSIBLE CAUSES	POSSIBLE CONSEQUENCES	SAFEGUARDS	RAC	RECOMMENDED ACTION(S)	REMARKS
1	Flow	No Flow	Valve closed	Pump deadhead	Engineering design; By-pass line	4D	None	None
			Pump failure	None	Back-up pumps	4D	None	None
			Pump off	None	Back-up pumps	4E	None	None
			Power failure	None	Back-up generator	4E	Ensure back-up is functional	None
			Line break	Spill of chemicals	Procedural control; Periodic inspection of lines; Monitored operations	3C	None	None
2	Flow	Low Flow	Pump reversed	Inadequate transfer; Tank vacuum not possible (Tank open to atmosphere)	Procedures require rotation check. Vent and line caps in place.	3C	Ensure operator training is adequate	None
			System obstruction	Inadequate transfer; Contamination	Procedures require rotation check. Vent and line caps in place.	3B	None	None
3	Pressure	High Pressure	Closed valve	Pump deadhead	Procedures; Training	3C	None	None
			System obstruction	Inadequate transfer; Contamination	Procedures require rotation check. Vent and line caps in pl	3B	None	None

Figure 14.4 *Partially complete HAZOP worksheet for the vapor degreaser system.*

in previous chapters). If the HAZOP and What-If exercises were conducted as stand-alone analyses, then a final written report should be developed to present all findings, recommendations, and conclusions. In this case, Table 14.4 provides a suggested outline of the content of such a report.

SUMMARY

The analytical techniques known as the HAZOP study and the What-If analysis have their roots in the chemical industry. Their specific utility, either when used together or separately, has been demonstrated time and again in the analysis of hypothetical failure events and scenarios in large and small-scale chemical production and processing facilities. However, there is no reason why the advantages of these analytical tools cannot be extended to industry in general.

TABLE 14.4   Sample Table of Contents for a HAZOP Final Report

Paragraph	Description
1.0	**INTRODUCTION**
1.1	GENERAL (general information about the project, process, or facility)
1.2	COMPANY POLICY (statement of company ES&H management policy)
1.3	PURPOSE & SCOPE (defines the objective and applicability of the HAZOP)
2.0	**PROCESS REVIEW TEAM**
2.1	AREAS REPRESENTED (identifies those work areas represented on the team)
2.2	TEAM MEMBERSHIP (provides name and contact information of team members)
3.0	**REFERENCE STANDARDS AND PROCEDURES**
	If there are only a few, they can be listed here. If the list is excessive, then an Attachment at the end of the Report is recommended
4.0	**DEFINITIONS**
	Provides essential definitions, such as HAZOP Study and What-If Analysis, here
5.0	**METHODOLOGY**
5.1	APPROACH (describes the approach taken to complete the analysis)
5.2	PARAMETERS (explains the operational parameters that were reviewed)
6.0	**ASSUMPTIONS**
	Describes the assumptions that were taken prior to the initiation of the analysis and upon which the analysis is based
7.0	**DRAWINGS**
	List and describe the drawings that were used to accomplish and/or support the analyses. Be sure to include drawing numbers and issue/revision dates.
8.0	**SUMMARY OF RECOMMENDATIONS**
8.1	WHAT-IF ANALYSIS RECOMMENDATIONS (summaries of results)
8.2	HAZOP ANALYSIS RECOMMENDATIONS (summaries of results)
9.0	**REVIEWER COMMENTS AND CONCLUSIONS**
9.1	GENERAL (provides general comments and conclusions for management)
9.2	SPECIFIC (if applicable, provide any specific comments and conclusions)
ATTACHMENT	**DESCRIPTION**
1	**ACRONYMS AND ABBREVIATIONS**
2	**WHAT-IF ANALYSIS WORKSHEETS**
3	**HAZOP REVIEW WORKSHEETS**
EXHIBITS	**DESCRIPTION**
1	**PHOTOGRAPHIC DOCUMENTATION**
2	**FACILITY SKETCHES OF PROCESS FLOW**

This chapter briefly described the use and application of both the HAZOP study and What-If analysis, when each can and should be performed during the various phases of the product or project life cycle, and how each can be used to evaluate the viability of a new or, more commonly, an existing system. The occupational safety and health practitioner should seriously consider using these techniques to ensure a more comprehensive analysis and overall understanding of the inherent safety of any system of process.

15

Special Use Analysis Techniques

INTRODUCTION

Although system safety, as a separate discipline, has existed for over 60 years, it is still subject to the ever-changing hazard reduction needs of modern-day industry. Its techniques and methods undergo constant modification and tailoring depending upon a specific hazard analysis requirement or set of requirements.

As technological advancements have continued to provide improvements over traditional methods of production, new types of hazard risk have also been introduced which are unique to these technologies. Therefore, to ensure a continued emphasis on the objective of risk reduction and/or control, certain system safety techniques have also been devised to address the particular types of hazard risk associated with new or expanding technologies.

To illustrate this point, this chapter will address two system safety analytical methods that have been developed as a result of technological improvements: The *Sneak Circuit Analysis* and *Software Hazard Analysis*. Each shall be briefly discussed here to demonstrate their applicability and utility in the practice of industrial safety and health.

Basic Guide to System Safety, Third Edition. Jeffrey W. Vincoli.
© 2014 John Wiley & Sons, Inc. Published 2014 by John Wiley & Sons, Inc.

SNEAK CIRCUIT ANALYSIS

The Sneak Circuit Analysis, or SCA, is a system safety analytical technique (also known as *sneak analysis*) used to identify and evaluate the different possible ways in which inherent system design characteristics can either

1. Permit an undesired function to occur,
2. Prevent a desired function from occurring, or
3. Adversely effect critical operational timing.

A "sneak" is a combination of conditions which cause an unexpected event. Such events are usually independent of hardware failure, meaning that they can occur during an operational mode when no system hardware failure is evident. Because these conditions can occur without any apparent of directly obvious cause, they may not be detected during systems tests. Hence the term "sneak" is particularly appropriate. Typically associated with analysis of electrical or electronic systems and other energy transfer systems (pneumatic, hydraulic, etc.), SCA has gained increasing popularity as more and more complex systems are being developed. Because of the insidious nature of sneak hazards, the occurrence is possible during all phases of the product or system life cycle.

From an industrial safety perspective, an excellent example where the SCA has particular applicability and utility is in the evaluation of the effectiveness of an organization's Lockout/Tagout (LOTO) program. Under the OSHA Standard for LOTO (i.e., The Control of Hazardous Energy), employers are required to verify the adequacy of their LOTO program at least annually to ensure that the minimum requirements established by OSHA are being followed [Ref: For General Industry— 29 CFR §1910.147(c)(6). For the Construction Industry—29 CFR §1926.417]. Use of an SCA to examine the effectiveness of the overall LOTO program in conjunction with other documented verification and inspection methodologies is an excellent approach to ensuring compliance with the OSHA inspection requirement. In addition, SCA can be used to verify the effectiveness and adequacy of a specific LOTO operation to ensure that all possible energy sources have been properly identified and isolated.

Types and Causes of Sneaks

Basically, the ways in which sneaks can occur and the potential results of these events can be grouped into four primary categories, as indicated in Table 15.1. Each type, if possibly applicable to a specific system, requires separate analysis. Since sneaks characteristically have no readily evident cause–effect relationships, particular attention to detail is essential in an SCA. For example, potentially devastating data or energy transfer may occur along an entirely unexpected path due to a minor error in system switching or circuitry design (a *sneak path*).

Sneaks occur due to a wide variety of causes, the most common of which include, but are not limited to, the following:

– Complete system overview is extremely difficult and, subsequently, potential sneak risks from design-phase errors are overlooked;

TABLE 15.1   Categories of "Sneaks" and Their Potential Effects

Sneak Circuit Analysis	
Category	Potential effect
Sneak Path	May inadvertently cause current or data to flow along an unexpected route or path leading to an increase in hazard risk and a possible fault event
Sneak Timing	May inadvertently cause current or data to flow at unexpected or unplanned times during system operation, which could result in system failure, damage, or loss
Sneak Indication	May cause a false, inaccurate, or otherwise confusing display of system operating conditions that could result in operator error
Sneak Label	Improperly labeled control sequences, operating instructions, hardware controls, etc., may lead to incorrect operator actions

- Improper assembly, connection, or command of devices and sensors during the production and/or operation phase;
- Inadvertent use of conflicting instruction labels, instructions, and/or operating procedures during the operational and even the disposal phases of the life cycle;
- As a result of changes or modifications to existing systems where all potential transfer paths or modes are not fully evaluated based upon the change.

SCA Input Requirements

To perform an SCA, the analyst requires access to all detailed design schematics and drawings, particularly those that integrate subsystems within a system. Wire lists and other such graphical information are also usual in the SCA. When specifications for particular components are available either from a vendor/supplier or generated within the design function, they should also be made available to the analyst during the SCA.

Although software hazard analysis is a separate system safety technique and is discussed later in this chapter as such, a review of compiler and/or assembly languages as well as any applicable system reference manuals and interface control specifications is advisable during the SCA. Sneak risks have been discovered due to improper or inappropriate software command initiatives.

As with all types of system safety analysis techniques discussed in this part, a complete description of the system, its intended purpose and design functions, as well as any operational flow diagrams must also be evaluated during the performance of an SCA. If the analyst is not entirely familiar with these system characteristics, the subsequent SCA will potentially be inaccurate, incomplete, and flawed.

While no particular worksheet is typically used during the performance of an SCA, an example of one is shown in Figure 15.1. The data collected here should be transferred to the Sneak Circuit Report for submission to management.

Advantages and Disadvantages of the SCA

There are obviously many possible benefits which result from utilization of the SCA technique, not the least of which are cost reduction and increased overall

		SNEAK CIRCUIT ANALYSIS		
PROGRAM: _____			DATE: _____	
ENGINEER: _____			PAGE: _____	
ITEM	EQUIPMENT EVALUATED	SNEAK EXPLANATION	POTENTIAL IMPACT	RECOMMENDED ACTIONS

Figure 15.1   *Sample sneak circuit analysis worksheet.*

system safety, reliability, and maintainability. Reduction in system development delays as well as fewer operational scheduling impacts are also potential advantages of the SCA.

In the area of cost reduction, for example, savings in the cost of the overall project will result due to the following considerations:

- Costs to perform an SCA are typically far less than those which could potentially occur as a result of undetected sneaks;
- Costs will most always be much less than that which will be required to fix a sneak detected much later in the product life cycle;
- Additional dollars will be saved if the SCA is performed in combination with other analyses.

An increase in the overall level of confidence in system safety and reliability will be realized through the use of the SCA as follows:

- The completed SCA provides an independent evaluation of the entire system (its hardware, software, and their interfaces within the system);
- The SCA will identify specific critical transfer and/or control problems that are typically missed during normal system functional testing;

– It provides a systematic, consistent, and thorough analytical review of data transfer/logic and/or current flow paths;

– The SCA method is conducive to the isolation of specific system faults which facilitate other analyses that may be performed.

In addition to those particular advantages highlighted above, the SCA will also reduce the time required to test a system by identifying specific problem areas ahead of time as well as those particular problems that may be created as a result of any engineering changes or system modifications.

Although there are not as many limitations as there are benefits associated with the use of the SCA, three primary concerns should be considered as follows:

– The SCA results do not usually provide an explanation of any definitive parameters for sneak occurrence. The potential hazard risk consequences associated with a given, identified sneak may vary dramatically depending upon circumstances and operational requirements.

– The SCA does not usually consider the effects of the many variable and potential environmental factors which may impose certain restrictions on the operation of a system.

– An SCA is not capable of validating the risk, if any, associated with software algorithms. The SCA will only perform software language and interface evaluations and cannot analyze the actual methods a system uses to solve a certain kind of problem. Hence, hazard risks associated with these methods may go undetected during analysis.

Software Hazard Analysis

Software Hazard Analysis, or SWHA, is a system safety analytical technique whose primary function is to systematically evaluate any potential faults in both operating system and applications software requirements, codes, and programs as they may effect overall system operation. The purpose of the SWHA is to ensure that safety specifications and related operational requirements are accurately and consistently translated into computer software programs. In this regard, the analysis will verify that specific operational safety criteria, such as fail-safe or fail-passive, have been properly assimilated into operational software. The SWHA will also identify and analyze those computer software programs, routines, or functions which may have direct control over, or indirect influence on, the safe operation of a given system. Also, in the operation of the computer software command function, there is a potential that the actual coded software may cause identified hazardous conditions to occur or inhibit a desired function thereby creating additional hazard potential.

Types of SWHA Techniques

Within the SWHA arena, a number of techniques have gained increasing popularity as automated system complexity progresses. Each type of SWHA has its use during the various phases of a product's life cycle, as described here.

The Software Preliminary Hazard Analysis Used to identify software program routines that are considered to be safety critical, this analysis is conducted *prior* to software program coding. To perform the analysis, the analyst should make reference to any available system specifications, interface documentation, functional flow diagrams, software flow charts, storage and file allocation specifications, and any other program descriptive information.

As a first-step evaluation of large, software intensive systems, it has the added advantage of providing for the separation of software interfaces into safety critical and safety noncritical functions which greatly facilitates the overall analysis effort. Obviously, proper performance of a software analysis in the preliminary stages will also reduce the potential cost of any subsequent analyses.

The major disadvantage is that, at the preliminary level, the analysis is not a structured technique; it is somewhat limited to only high level functions and, thus, additional analyses of complex systems are often required, especially where safety critical software functions have been identified.

Software Fault Hazard Analysis Similar in concept and structure to the system hazard analysis (SHA) which is conducted on system hardware, the software fault hazard analysis will analyze and evaluate a computer software program to identify critical areas in the programming which may contribute to or directly cause a hazard risk. Such risks may be due to an undetected hardware failure or incorrect inputs into the operation of the system software. The software "FHA" will also attempt to uncover any probable errors that can possibly develop in the software after system activation.

Since input and/or output variables typically fail in a discrete manner, the evaluation of any single-point failure effects upon the software program, the computer hardware, and/or the system can be accomplished through the software FHA. The drawback here is that the specific effect of any such failure may be somewhat difficult to define since they are typically a function of the actual operational state of the computer at the exact time the failure occurs. Hence, even evaluation of hypothetical scenarios could be a monumental task since so many possible variables are bound to exist. Also, the software FHA can be an extremely lengthy and quite tedious process that may or may not yield any significant results. A decision to proceed with such an effort must therefore be weighed against the anticipated benefits which are expected upon its completion.

Software Fault Tree ("Soft Trees") The soft tree technique is used to determine what software event, failure, or combination of each will result in a real or hypothetical loss event (a *top event*). This top-down analytical approach, which assumes a problem and then evaluates effecting conditions backward to determine causal factors, also takes into consideration any influencing environmental factors. It is primarily concerned with the analysis of any hardware/software interfaces that deal directly with the operation of mechanical components.

Because it is a fault tree technique, the soft tree provides a graphic illustration of the potential effect of multiple and/or simultaneous failures. The tree method also allows for the quantification of analysis results, if desired. The primary disadvantage here is that, when using a fault tree method for hypothetical analysis of a loss event, the analysis is subjective since the analyst must choose which events to evaluate. The risk is that not all possible events will be considered. Because the soft tree is also a costly exercise, certain decisions must be made as to the extent of the analysis and, hence, the likelihood that potential loss events will be overlooked is further increased. Also, the soft tree does not consider timing in its evaluation, only simultaneous failure events, and it only evaluates what is there to evaluate and does not consider the potential impact of unintentional omissions which may also adversely impact overall system safety.

Emulation Analysis   This technique determines the ability of established software programming to detect specific hardware and/or software faults purposely introduced into the microprocessing system. Usually, output results from the tested system are compared to that of a controlled, uninfected system to determine if all faults were properly detected. This method allows for the quantification of faults detected in microprocessor or program codes and provides a method for bit manipulation of software programming.

   Timing in emulation analysis is much slower than that of the actual program and, therefore, the results may not accurately predict true bit coverage under real-time operating conditions. The method is also quite costly to perform.

Software System Hazard Analysis   Conducted similar to a hardware SHA, this method analyzes software functional processing steps to determine if they may have any particular hazardous effect on the system. The analysis utilizes a hazard-risk index to illustrate the severity of each potential failure. The main advantage to this method is in its ability to positively identify safety critical hardware and software functions as well as consider the effect of the human element in system software operations. The results of the software SHA, which identifies single-point failures or errors within a system, can often be used to assist in the development of a software fault tree analysis or, to some degree, a system FMEA. However, as with the other various SWHA techniques briefly described above, this method is also time-consuming and costly to perform.

SUMMARY

As the practice of system safety moves into its seventh decade of existence, special use analysis methods and techniques have been developed due to the steady increase in system complexity and acceleration in technological advancements. These changes have brought with them new concerns over the adequacy of existing hazard reduction and control techniques. Automated systems that incorporate complicated data and energy flow transfer paths, as well as those who's correct operation depends primarily upon proper software programming, can impose significant risk of system hazards that would otherwise go undetected during normal system testing.

The system safety analysis techniques known separately as *sneak circuit analysis* and *software safety analysis* have been developed in an effort to address these concerns over system safety and reliability assurance. While there are various types of sneak hazards which can be identified by analysis, and a variety of software hazard analysis techniques are commonly used, each method is primarily concerned with the same essential objective explained throughout this text: hazard risk elimination or reduction to acceptable levels.

Epilogue

This second edition of a *Basic Guide to System Safety* has been designed to provide the reader with a fundamental understanding of the system safety discipline, the assessment of risk, the hazard analysis process, and some of the common tools and techniques that can be used to determine levels of hazard risk. Numerous examples have been developed throughout the text in an attempt to demonstrate the applicability of system safety engineering and analysis in the practice of the industrial safety and health professional.

In fact, the primary objective of the text has been to impart a basic level of appreciation for the value and utility of a working system safety program in the occupational or industrial safety and health arena. In order to accomplish this, it is considered crucial that the reader first have a clear understanding of how system safety developed as a necessary subdiscipline of systems engineering during the early missile and space programs of the United States Air Force. Acknowledgment of these beginnings will help the reader further appreciate the ability of proper system safety engineering and analysis to ensure maximum reduction of hazard risk in a given system, product, program, or service.

Therefore, Part I of this text focused primarily on the development of system safety, its military connections, the importance of including system safety requirements in contract acquisitions, the criticality of obtaining management commitment in support of the system safety effort, the process of risk analysis and assessment, probability theory and statistical analysis as they relate to system safety, and, perhaps of most value, how the fundamental principals of system safety are closely related to those of occupational safety and health management.

Basic Guide to System Safety, Third Edition. Jeffrey W. Vincoli.
© 2014 John Wiley & Sons, Inc. Published 2014 by John Wiley & Sons, Inc.

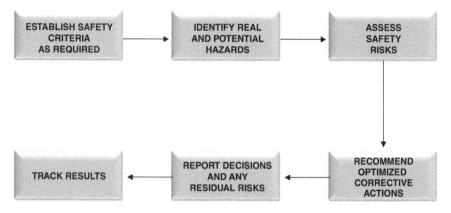

Figure E.1 *Summary of the system safety process.*

In Part II, the reader was exposed to a variety of the most common tools and techniques currently used in the system safety profession. It is hoped that the numerous examples provided will assist in developing an appreciation for system safety analysis in the evaluation of risk, no matter how complex or simple the system may be. Although these various examples did not constitute complete and detailed analyses, it is presumed that enough information has been presented to ensure a basic understanding of common system safety analysis techniques and methods.

Figure E.1 represents the simple flow of the system safety process and provides a graphic summary of the materials presented in this text. This flow shows the typical functions of the system safety life cycle.

Finally, upon completion of this text, the reader should be in a position to make an informed decision to further pursue the field of system safety engineering, analysis, and/or management based upon their individual needs. Appendix A provides a brief listing of respectable sources where further, more detailed information and training may be obtained.

Whatever your decision may be, it will help to remember that ancient Chinese proverb: ... *if you don't know where you are going, then any road will take you there!* A properly implemented and managed system safety program can provide an excellent "road map" to help individuals, organizations, and entire corporations find their way to a safe, productive, and profitable destination while ensuring the lowest possible level of acceptable risk with a maximum return on investment.

Appendix A

Sources of Additional Information/Training

The following is a compilation of sources where the interested reader may obtain additional information and/or training in the area of system safety engineering and/or management. There are, of course, many more excellent references available to the system safety practitioner. However, those listed here will provide the reader of this *Basic Guide to System Safety* additional information presented at the next technical level.

Professional Organizations:

1. *The International System Safety Society (SSS)*
   P.O. Box 70
   Unionville, VA 22567-0070
   (504) 854-8630
   http://www.system-safety.org/

An international, nonprofit organization in the United States and other countries around the world. The SSS is dedicated to the safety of systems, products, and services. Originally organized in 1962, it was incorporated in 1973 and has its head-quarters in the Washington, DC area. Active chapters are organized to promote the system safety philosophy and further professional development. Their journal, *Hazard Prevention*, is published on a quarterly basis and features articles on current developments in the system safety profession.

Basic Guide to System Safety, Third Edition. Jeffrey W. Vincoli.
© 2014 John Wiley & Sons, Inc. Published 2014 by John Wiley & Sons, Inc.

2. *American Society of Safety Engineers (ASSE)*
   1800 East Oakton Street
   Des Plaines, IL 60018-2187
   (847) 699-2929
   http://www.asse.org/

The ASSE is an international organization with over 34,000 members (2012) in the United States and in selected countries such as England and Saudi Arabia where numerous American Safety Professionals work and live. Organized in 1911 and incorporated in 1915, it is one of the oldest sustaining professional safety membership organizations in the United States. Through its many Counsels, the ASSE is an excellent source of information on a wide variety of safety and health topics, including system safety. Their monthly journal, *Professional Safety*, often includes articles on the subject of system safety analysis.

3. *Board of Certified Safety Professional (BCSP)*
   2301 West Bradley Avenue
   Champaign, IL 61821
   (217) 359-9263
   http://www.bcsp.org/

The BCSP is involved in the development, coordination, implementation, administration, and maintenance of professional certification examinations to individual practitioners of various safety and health disciplines, including system safety. They offer examination sessions throughout the year at specific, approved locations across the country. The BCSP can provide additional detailed information concerning certification requirements.

4. *Association of Computing Machinery (ACM)*
   2 Penn Plaza, Suite 701
   New York, NY 10121-0701
   (800) 342-6626
   http://www.acm.org/

ACM is an excellent reference source on the current developments in software safety analysis. Their publication, *Computing Machinery,* often features articles from top industry leaders on the subject of software safety. ACM is the world's largest educational and scientific computing society that delivers resources to advance computing as a science and a profession. ACM provides the computing field's premier Digital Library and serves its members and the computing profession with leading-edge publications, conferences, and career resources.

Publications: Many excellent publications exist on the subject of system safety. The following are excellent, highly recommended reference sources for intermediate and advanced system safety studies:

1. Title: *Job Hazard Analysis: A Guide for Voluntary Compliance and Beyond*
    Author/Year: J. Roughfton and N. Crutchfield/2008
    Publisher: Elsevier, Inc.
                  Oxford, England OX2 8DP

2. Title: *Hazard Analysis Techniques for System Safety*
    Author/Year: C. Ericson/2005
    Publisher: John Wiley and Sons, Inc.
                  Hoboken, NJ 07030

3. Title: *System Safety for the 21st Century*
    Author/Year: R. Stephans/2004
    Publisher: John Wiley and Sons, Inc.
                  Hoboken, NJ 07030

4. Title: *System Safety 2000*
    Author/Year: J. Stephenson/1991
    Publisher: Van Nostrand Reinhold
                  New York, NY 10003

5. Title: *System Safety Engineering and Management*
    Author/Year: H. E. Roland and B. Moriarty/1991
    Publisher: John Wiley & Sons
                  New York, NY 10158

6. Title: *Safety and Reliability in System Design*
    Author/Year: M. Larson and S. Hann/Undated
    Publisher: Ginn Press
                  Needham Heights, MA 02194

7. Title: *System Safety Technology and Application*
    Author/Year: S. W. Malasky/1982
    Publisher: Garland STPM Press
                  New York, NY 10102

Training Courses and Seminars:

1. Sponsor: SoHar, Inc.
    Address: 5721 West Slauson Avenue, Suite 140
               Culver City, CA 90230
               310-338-0990
               http://www.sohar.com/index.html

SoHaR's system safety training program is an integral part of the company's consulting services and as such draws the expertise and experience of the best professionals. Courses are developed and designed to reflect current practices and include both hands-on material and theoretical explanations. The goal is to provide not only formulas to solve problems but also the know-how to deal with challenges and cases that may be outside of the mainstream practice.

2. Sponsor: American Society of Safety Engineers
   Address: 1800 East Oakton Avenue
               Des Plaines, IL 60018-2187
               (847) 699-2929
               http://www.asse.org/

The American Society of Safety Engineers ("ASSE") is a national, professional organization with over 35,000 members (2012) practicing in all aspects of the safety profession, including system safety. The organization sponsors numerous training seminars and a yearly Professional Development Conference. A number of excellent system safety courses are included in their extensive listing of available training courses.

3. Sponsor: HCRQ, Inc.
   Address: P.O. Box 264
               Williamsburg, VA 23187
               757-564-7703
               http://www.asse.org/

The company has extensive expertise in system safety (dating to 1988) and provides consulting services in all safety-critical sectors. They provide both training courses and webinars in system safety and are fully conversant with all system safety standards. They can develop specific training to meet client needs as well as provide standard training in all areas of system safety analysis.

4. Sponsor: University of Washington
               Professional and Continuing Education
   Address: 1410 NE Campus Parkway
               Seattle, WA 98195
               206-543-2544
               http://www.washington.edu/

In a course entitled *System Safety Management*, the University of Washington offers an in-depth review of the management tasks appropriate for each phase of a system's life cycle, including the various life cycle phases of military and nonmilitary systems and facilities, and an overview of the analytical and mathematical theory necessary to perform system safety engineering tasks.

5. Sponsor: George Washington University
    Address: 2121 Eye Street, NW
                  Washington, DC 20052
                  202-994-1000
                  http://www.gwu.edu/

The university's continuing education program offers several courses in system safety analysis and system reliability techniques. Most courses are taught in Washington, DC with some being presented at alternate locations in the United States.

6. Sponsor: University of Southern California
                  Institute of Safety and Systems Management
    Address: University Park Campus
                  Los Angeles, CA 90089-0021
                  213-740-2311
                  http://www.usc.edu/

Although most courses are primarily geared toward the aviation industry, the university does offer an exceptional system safety course as well as a software safety seminar.

7. Sponsor: Safety Links, Inc.
    Address: 687 South Bluford Avenue
                  Ocoee, FL 34734
    URL: http://www.safetylinks.net/

Safety Links was founded in 1997 as a safety training provider. As the organization grew, their focus shifted to safety consultation and then finally to business management solutions. At that time, their clientele ranged from small manufacturing firms to large construction companies. In 2003, Safety Links relocated to Orlando, Florida, and has built strong relationships with a core group of clients and has progressively grown as a result of referrals. Their services range across the broad spectrum of consulting and training in the areas of occupational safety and health (to include job hazard analyses).

Acronyms and Abbreviations

In the practice of system safety engineering and management, as well as in the safety and health profession in general, numerous abbreviations and acronyms are used quite regularly. The following is a reference listing of those most frequently used or encountered, either in this text or in the system safety and/or safety and health disciplines in general.

ACM	Association of Computing Machinery
AFOSH	Air Force Occupational Safety and Health
AFR	Air Force Regulation
AMA	American Medical Association
ANSI	American National Standards Institute
ARAR	Accident Risk Assessment Report
ASME	American Society of Mechanical Engineers
ASSE	American Society of Safety Engineers
BCSP	Board of Certified Safety Professionals
CASCA	Computer-Aided Sneak Circuit Analysis
CBA	Cost-Benefit Analysis
CGA	Compressed Gas Association
CCFA	Common Cause Failure Analysis
CDRL	Contract Data Requirements List
CEA	Cause and Effect Analysis

Basic Guide to System Safety, Third Edition. Jeffrey W. Vincoli.
© 2014 John Wiley & Sons, Inc. Published 2014 by John Wiley & Sons, Inc.

CFC	Chlorofluorocarbon
CFR	Code of Federal Regulations
CPSC	Consumer Product Safety Commission
CSPF	Critical Single-Point Failure
CSP	Certified Safety Professional
DID	Data Item Description
DOD	Department of Defense
DOE	Department of Energy
DOT	Department of Transportation
ECP	Engineering Change Proposal
EO	Engineering Order
EPA	Environmental Protection Agency
ETA	Event Tree Analysis
ETBA	Energy Trace and Barrier Analysis
EVA	Extreme Value Analysis
FAA	Federal Aviation Administration
FAR	Federal Aviation Regulation
	Federal Acquisition Regulation
FHA	Fault (or Functional) Hazard Analysis
FM	Factory Mutual
FMEA	Failure Mode and Effect Analysis
FMECA	Failure Mode and Effect Criticality Analysis
FR	Federal Register
FTA	Fault Tree Analysis
GHA	Gross Hazard Analysis
GP	Government Practice
HA	Hazard Analysis
HAZOP	Hazard and Operability Studies
HFS	Human Factors Society
HPSSC	Health Physics Society Standards Committee
HTI	Hand Tools Institute
IEEE	Institute of Electrical and Electronics Engineering
IES	Illuminating Engineering Society
IME	Institute of Makers of Explosives
ISEA	Industrial Safety Equipment Association
ISSPP	Integrated System Safety Program Plan
JSA	Job Safety Analysis
JHA	Job Hazard Analysis
LSC	Life Safety Code
MIL-STD	Military Standard
MORT	Management Oversight and Risk Tree
NASA	National Aeronautics and Space Administration
NASHP	National Association of Safety and Health Professionals
NAVOSH	Navy Occupational Safety and Health

NBS	National Bureau of Standards
NEC	National Electrical Code
NEMA	National Electrical Manufacturers Association
NFPA	National Fire Protection Association
NHB	NASA Handbook
NIOSH	National Institute for Occupational Safety and Health
NSC	National Safety Council
NSTS	National Space Transportation System
NTSB	National Transportation Safety Board
OHA	Operating Hazard Analysis
OHHA	Occupational Health Hazard Assessment
ORI	Operational Readiness Inspection
ORR	Operational Readiness Review
O&SHA	Operating and Support Hazard Analysis
OSHA	Occupational Safety and Health Administration
PET	Project Evaluation Tree
PHA	Preliminary Hazard Analysis
PHL	Preliminary Hazard List
PO	Purchase Order
PR	Problem Report
	Purchase Request
PTI	Power Tool Institute
RAC	Risk Assessment Code
RFP	Request for Proposal
RIA	Robotic Industries Association
RNSA	Random Number Simulation Analysis
SAE	Society of Automotive Engineers
SCA	Sneak Circuit Analysis
SPF	Single-Point Failure
SIC	Standard Industry Classification
SOP	Standard Operating Procedure
SOW	Statement of Work
SHA	System Hazard Analysis
SSA	System Safety Analysis
SSDC	System Safety Development Center
SSHA	Subsystem Hazard Analysis
SSPP	System Safety Program Plan
SSS	System Safety Society
SSWG	System Safety Working Group
SWHA	Software Hazard Analysis
TCP	Task Change Proposal
THERP	Technique for Human Error Rate Prediction
TLA	Time-Loss Analysis
TORAR	Technical Operations Risk Assessment Report

TQM	Total Quality Management
TQMS	Total Quality Management System
UL	Underwriter's Laboratories
WHO	World Health Organization
WSO	World Safety Organization
ZA	Zonal Analysis

Glossary of Terms

The following are definitions for many of the terms which appear in this text or are encountered in the practice of system safety. When appropriate, the source of an individual definition is referenced parenthetically at the end of the definition.

Absolute Pressure: Pressure measured with respect to zero pressure or a vacuum. It is equal to the sum of a pressure gauge reading and the atmospheric pressure at the measurement location.

Absolute Temperature: Temperature based on an absolute scale expressed in either degrees Kelvin or degrees Rankine corresponding, respectively, to the centigrade or Fahrenheit scales. Degrees Kelvin are obtained by adding 273 to the centigrade temperature or subtracting the centigrade temperature from 273 if below 0°C. Degrees Rankine are obtained by algebraically adding the Fahrenheit reading to 460. Zero K is equal to −273°C and zero R is equal to −459.69°F.

Acceleration: A vector representing the rate of change of velocity with time.

Acceleration Power: Measured in kilowatts. Pulse power obtainable from a battery used to accelerate a vehicle. This is based on a constant current pulse for 30 seconds at no less than 2/3 of the maximum open-circuit voltage, at 80% depth-of-discharge relative to the battery's rated capacity and at 20°C ambient temperature.

Acceptability: With regard to the use of instruments, the willingness of personnel to use an instrument when considering its characteristics, such as weight, noise, response time, drift, portability, reliability, interference effects, etc.

Basic Guide to System Safety, Third Edition. Jeffrey W. Vincoli.
© 2014 John Wiley & Sons, Inc. Published 2014 by John Wiley & Sons, Inc.

Acceptable Risk: The residual risk that remains after all possible control measures have been implemented that is deemed acceptable by the party or parties that are exposed to the risk (e.g., management, employees, the public, the government).

Accident: An unwanted event resulting from the occurrence of one or more fault incidents which have a negative impact on a system, product, equipment, or personnel.

Accident Analysis: A concerted, organized, methodical, planned process of examination and evaluation of all evidence and records identified during investigation of accidents.

Accident Investigation: A detailed and methodical effort to collect and interpret facts related to an individual accident, conducted to identify the causes and develop control measures to prevent recurrence; a systematic look at the nature and extent of the accident, the risks taken, and loss(es) involved; an inquiry as to how and why the accident event occurred.

Accident Phases: In an accident investigation, when evaluating the sequence of events that resulted in an accident, the events are divided into three phases or categories: *precontact* (before the accident), *contact* (the accident), and *postcontact* (after the accident). Analysis of the events occurring in each phase facilitates the identification of loss-inducing activities and conditions. Also referred to as the three stages of loss control.

Accident Potential: A situation comprised of human behaviors and/or physical conditions having a probability of resulting in an accident.

Accident Risk: A measure of vulnerability to loss, damage, or injury caused by a dangerous element or factor (MIL-STD-1574).

Accident Risk Assessment: A written evaluation of those hazards associated with the operation of a given facility, including any equipment or hardware used in that facility. A determination of the accident potential and an explanation of control measures are also provided.

Accident Risk Factor: A dangerous element of a system, event, process, or activity including casual factors such as design or programming deficiency, component malfunction, human error or environment, which can propagate a hazard into an accident if adequate controls are not effectively applied (MIL-STD-1574).

Accident Sources: Accidents generally involve one or all of five elements: people, equipment, material, procedures, and the work environment, each of which must interact for successful business operations. However, when something unplanned and undesired occurs within either of these elements, there is usually some adverse effect on any one or all of the other elements, which if allowed to continue uncorrected, could lead to an incident or accident and subsequent loss.

Act of God: An act occasioned by an unanticipated grave natural disaster or other natural phenomenon of exceptional, inevitable, and irresistible character, the effects of which could not have been prevented or avoided by the exercise of due care or foresight.

Active Restraint: A restraining device which has a positive locking feature and requires no action by an individual to be held in place. An example would be a seat belt system in an automobile at the time of collision.

Active Safety Measure: Any means of implementing safety precautions which requires an individual to take some action, such as reading or comprehending. An example would be a warning sign indicating an unsafe or hazardous condition.

Administrative Control: A measure initiated to reduce worker exposure to various stresses in the work environment. An example is limiting the amount of time an employee can work around health hazards.

Air Sampling: The collection of samples of air to determine the presence of and the concentration of a contaminant, such as a chemical, aerosol, radioactive material, airborne microorganism, or other substance by analyzing the collected sample to determine the amount present and calculating the concentration based on the sample volume.

Airborne Particulates: Total suspended particulate matter found in the atmosphere as solid particles or liquid droplets. The chemical composition of particulates varies widely, depending on location and time of year. Airborne particulates include windblown dust, emissions from industrial process, smoke from burning of wood and coal, and exhaust of motor vehicles.

Alarm: An indicator that some condition exists which may or will require human action to correct in order to prevent loss of life, property, or equipment.

Amplitude: The instantaneous deviation or displacement from some baseline. The peak-to-peak difference, maximum value, or averaged value of a signal.

Analysis: A study or evaluation, usually performed to determine the current status of a given system or process. It will often utilize established standards or operating criteria as a baseline for comparison.

As-Built Plan: A drawing which covers property boundaries, streets bordering the site and building layout, and provides accurate scale and a north arrow.

Assembly: A combination of multiple components or parts grouped together to perform a single function or a specific set of functions within a system or subsystem.

Assessment: An evaluation or examination of a specific area of concern, such as a program, policy, or procedural assessment.

Assistive Device: Any tool which either enables or enhances human–machine interaction for an individual with a physical handicap.

Audit: A detailed and systematic inspection or review of an occupational health and safety program, environmental program, financial operating program, or some other program, to determine compliance with company policies, practices, and procedures, as well as the regulations that are applicable to the operations and work being performed.

Barrier: A control (device, mechanism, structure, sign, etc.) intended to prevent the transfer of energy from one element of a system to another. Any object, individual, or structure which impeded progress toward a goal or which prevents entry to a region for safety reasons.

Basic Event: As pertains to fault tree analysis (FTA) and/or the Management Oversight and Risk Tree (MORT), a root fault event or the first in the process to have occurred that requires no further development or analysis. Represented graphically as a circle.

Bathtub Curve: A graphical representation of the life cycle of products, systems, or individual components in terms of frequency of failures relative to periods of usefulness. In system safety, it is also known as a *reliability curve.*

Behavior: Any rating scale developed to evaluate individual behavioral patterns.

Bench Test: A small-scale test or study used to determine whether a technology is suitable for a particular application.

Benchmark: A thoroughly documented reference value or standard or measurement against which performance, response, or other characteristics may be compared with confidence.

Binomial Distribution: A distribution of data or results describing probabilities of the outcome of trials that can have one or two mutually exclusive results (e.g., exposure above or below a permissible exposure limit or "PEL"). This theoretically discrete probability distribution for a binomial random variable is represented as:

$$P = n/r \, p^n (1 - p)^{n-r},$$

where n = total number of outcomes, r = number of successful outcomes, (n/r) = number of combinations of n outcomes, taken r at a time.

Used to approximate the normal distribution for large sample sizes.

Blanking: The absolute closure of a pipe, line, or duct by the fastening of a solid plate (such as a spectacle blind or skillet blind) that completely covers the bore and that is capable of withstanding the maximum pressure of the pipe, line, or duct with no leakage beyond the plate. Also called *blinding.*

British Thermal Unit (BTU): The amount of energy required to raise the temperature of 1 pound of water 1 degree Fahrenheit (°F) at or near 39.2°F and 1 atmosphere of pressure. One British Thermal Unit (BTU) is about equal to the heat given off by a blue-tip match.

Carcinogen: A substance known to cause cancer in humans and animals representing a broad range of organic and inorganic chemicals, hormones, immunosuppresants, and solid-state materials.

Carelessness: That behavior or mental functioning which does not exhibit adequate attention or concern for the task being performed.

Carpal Tunnel: An internal passage in the wrist between the extensor retinaculum and the carpal bones through which the median nerve, finger flexor tendons, and blood vessels pass from the arm to the hand.

Carpal Tunnel Syndrome: A cumulative trauma disorder (CTD) often associated with activities involving flexing or extending the wrists or repeated force on the base of the palm and wrist. The *carpal tunnel* is an opening in the wrist under the carpal ligament on the palmar side of the carpal bones in the wrist. The median nerve, the finger flexor tendons, and blood vessels all pass through this tunnel. Overuse of the tendons can cause them to become inflamed and swollen, creating pressure against the adjacent median nerve and resulting in CTS. Symptoms include tingling, pain, or numbness in the thumb and first three fingers.

Catastrophic Event: An occurrence, subsequent to the introduction of a hazard or set of hazards into a system, that results in a level of injury, damage, or loss of such severe magnitude that quick or total recovery would be highly improbable (e.g., death, crippling injuries, total system loss, irreplaceable property or equipment loss or damage). The parameters for this categorization are usually established by management in the System Safety Program Plan, or other policy-making documentation.

Catastrophic Release: According to OSHA, a major uncontrolled emission, fire, or explosion, involving one or more highly hazardous chemicals, that presents serious danger to employees in the workplace.

Causal Association: Having a demonstrable connection between the occurrence of some factor and an incident, where the presence of that factor will increase the probability and the absence of that factor will decrease the probability of that incident.

Causal Factors: A combination of simultaneous or sequential circumstances which contribute directly or indirectly to an accident, occupational disease, or other effect.

Cause: In Safety, an event, situation, or condition which results, or could result *(potential cause)*, directly or indirectly, in an accident or incident. Each separate antecedent of an event. Something that proceeds and brings about an effect or result.

Cause-Effect Diagram: A graphical display of the causes linked to an effect.

Closed-Loop System: Any type of system in which the output or some derivative of the output from the system is directed back into the system itself. Synonymous with *feedback control loop.*

Coefficient: A number by which one value is to be multiplied in order to give another value, or a number that indicates the range of an effect produced under certain conditions.

Color Coding: The use of multiple colors for easier, more rapid visual identification, access, and/or processing of groups of organized materials.

Combustible: Capable of being ignited with resultant burning or explosion.

Common Cause Failure Analysis: A system safety analytical technique (also known as *common cause analysis*) used primarily in the evaluation of multiple failures that have the occurrence of a single event as a common causal factor.

Component: A functional part of a subsystem or equipment which is essential to operational completeness of the subsystem or equipment and which may consist of a combination of parts, assemblies, accessories, and attachments (Larson and Hann 1989).

Concept Phase: That portion of a system's, product's, or other yet to be developed program's life cycle during which ideas are first conceptualized; precedes the design phase.

Concurrent Causes: Causes acting contemporaneously and together causing injury, which would not have resulted in the absence of either. Two distinct causes operating at the same time to produce a given result, which might be produced by either, are considered concurrent causes. However, two distinct causes, successive and unrelated in an operation, cannot be concurring, and one will be regarded as the proximate and efficient and responsible cause, and the other will be regarded as the remote cause.

Conditional Event: As pertains to fault tree analysis (FTA) and/or the Management Oversight and Risk Tree (MORT), an occurrence that, based upon its own unique characteristics, imposes conditions or exclusions on the occurrence of other events in the fault path. Represented graphically as an oval. See also *Exclusive Event*.

Contingency Analysis: An analysis performed to identify what abnormal situations, errors, or malfunctions, a system may develop or encounter to improve system performance or establish what special human responses may be required under those circumstances.

Contributory Event: As pertains to fault tree analysis (FTA) and/or the Management Oversight and Risk Tree (MORT), an event which significantly influences the outcome of the top or primary event. Also known as a main event or secondary event.

Coolant: A liquid or gas used to reduce the heat generated by power production in nuclear reactors, electric generators, various industrial and mechanical processes, and automobile engines.

Cost-Benefit Analysis: A system safety analytical technique used to evaluate various possible courses of action with respect to the costs that are incurred compared to the benefit of the results.

Credible Failure: Any failure that can physically occur without violating any scientific law (Larson and Hann 1989).

Critical Condition: The most severe environmental condition in terms of loads, pressures, and temperatures, or combinations thereof. Imposed on structures, systems, subsystems, and components during service life.

Critical Event: An occurrence, subsequent to the introduction of a hazard or set of hazards into a system, that results in a level of injury, damage, or loss of a magnitude that quick or total recovery would be possible, although extremely difficult (e.g., personnel injuries, partial system loss, property or equipment damage). The parameters for this categorization are usually established by management in the System Safety Program Plan, or other policy-making documentation.

Critical Incident Method: A performance appraisal technique for either a system or employee. For a system: the process of gathering data by asking the users of that system to describe significant incidents, according to some established criteria. For an employee: the maintenance of a log documenting both favorable and unfavorable behaviors exhibited during an evaluation period. Synonymous with *critical incident technique.*

Criticality: A scale or ranking of the possible types of failures in a system as to the importance of continued functioning of that system.

Crossover Analysis: An evaluation for costing purposes of what alternative work methodologies should be used for different production levels.

Cumulative Error: An error whose sum does not converge to zero as the number of samples increases.

Cut-Set: As pertains to fault tree analysis (FTA) and/or the Management Oversight and Risk Tree (MORT), a defined set of events, under the top event, that can be isolated from the remainder of the fault tree and examined as contributory to the occurrence of the top or primary event.

Damage: The partial or total loss of hardware caused by component failure; exposure to heat, fire, or other environments; human errors; or other inadvertent events or conditions (MIL-STD-882).

Danger: Term of warning applied to a condition, operation, or situation that has the potential for physical harm to personnel and/or damage to property.

Dangerous Condition: One in which there exists a substantial and probable risk of injury and/or property damage. The risk may be imminent or merely possible when such a condition exists.

Dead Man Control: A device requiring a constant force of a minimum magnitude applied to the device for operating a piece of equipment, and having a default mode which turns off or stops the equipment if that force is not applied.

Defect: Substandard physical condition, either inherent in the material or created through another action or event.

Degreaser: A chemical agent, usually a solvent that is used to remove grease and oil from machinery. Because these chemicals will also remove the protective layer of oil on human skin, their use without protection can result in dermatitis.

Design: The process of developing the requirements, structure, dimensions, tolerances, and materials to be used for an entity.

Design Safety Factor: A factor used to account for uncertainties in material properties and analysis procedures. It is often referred to as *design factor of safety* or simply *safety factor.*

Deviation: An alternate method of compliance with the intent of specific requirements (MIL-STD-1574A). A departure from established or usual conduct or ideology. (2) The amount by which a score or other measure differs from the mean or other descriptive statistic.

Electric System: Physically connected generation, transmission, and distribution facilities operated as an integrated unit under one central management or operating supervision.

Electric Component: A component such as a switch, fuse, resistor, wire, capacitor, or diode in an electrical system.

Emergency Procedure: An action plan to be implemented in the event of an emergency. It typically describes, as a minimum, roles and responsibilities, types of emergency situations to be expected, emergency notification and/or communication procedures, public relations procedures during an emergency, and any other contingency plans applicable to the facility and its processes.

Emergency Stop: A push button, switch, or other control device installed in or on a piece of equipment which is capable of quickly cutting power to that equipment in an emergency.

Empirical Distribution: A distribution of sampled events or data.

Empirical Probability: When many possible outcomes can result, including a desired outcome, the probability of occurrence of such outcomes is referred to as empirical and requires statistical evaluation to determine the likelihood of expected results based upon past performance.

Empirical Workplace Design: The evolutionary design of the working environment based on a combination of human factors engineering and experience.

Energy Trace and Barrier Analysis: A system safety analytical technique used to evaluate the flow of energy through a system and analyze the effectiveness of existing barriers within the system which are intended to prevent unwanted transfers of that energy flow.

Engineering: A discipline in which knowledge of the mathematical and natural sciences, gained by some combination of education, training, and practical experience, is integrated with various natural materials and forces to shape the environment.

Engineering Controls: Measures taken to prevent or minimize hazard exposure through the application of controls such as improved ventilation, noise reduction techniques, chemical substitution, equipment and facility modifications, etc.

Ergonomics: The scientific study and analysis of the human, machine, and/or working environment interface and an investigation of those elements in the system that effect optimum human performance on a given task or set of tasks.

Error: The difference between the true or actual value to be measured and the value to be measured and the value indicated by the measuring system. Any deviation of an observed value from the true value. (2) An inappropriate response by a system, whether of commission, omission, inadequacy, or timing.

Event Tree: A graphical depiction of system or operational events as they are related to the top event or failure condition.

Event Tree Analysis: A system safety analysis method, similar to fault tree analysis, used to examine different system or operational responses to various positive or negative conditions which occur during system operation.

Exclusive Event: As pertains to fault tree analysis (FTA) and/or the Management Oversight and Risk Tree (MORT), a conditional event which places specific restrictions upon the occurrence of other events. Represented graphically as an oval. See also *Conditional Event.*

Explosion: A rapid build-up and release of pressure caused by chemical reaction or by an overpressurization within a confined space leading to a massive rupture of the pressurized container.

Extreme Value Projection: In system safety, a risk projection technique used to provide information about potential losses (i.e., in the future) that are more severe than those occurring in the past.

Fail: To fall short; be unsuccessful or deficient.

Fail Operational: A design characteristic which allows continued operation of a system or subsystem despite a discrete failure.

Fail Operational, Fail Safe: A fail operational design which also remains acceptably safe.

Fail Passive: A system or component design feature that, under failure conditions, will have no effect on the operation of the overall system.

Fail Safe: A system or component design feature that, under failure conditions, will permit the failed component or system to revert to a safe mode and not present an unacceptable hazard risk or flow of energy due to the failure condition.

Failure: The inability of a component or system to perform its designed function within specified limits.

Failure Assessment: The process in which the cause, effect, responsibility, and cost of a failure are determined and reported.

Failure Condition: As pertains to fault tree analysis (FTA) and/or the Management Oversight and Risk Tree (MORT), the top event, or that primary event subject to a failure analysis through an event tree.

Failure Mode and Effect Analysis: An in-depth analysis of possible failures and their effects related to system functions *(functional FMEA)* or system hardware and components *(hardware FMEA)*.

Failure Tolerance: The ability of a system to experience one or more failures and still maintain some functional capability.

Fatigue (Structural): The progressive localized permanent structural change that occurs in a material subjected to constant or variable amplitude loads at stresses having a maximum value less than the ultimate strength of the material.

Fault (or Functional) Hazard Analysis: A system safety analysis method, usually an extension of the *failure mode and effect analysis* that evaluates the overall effect of functional failures on other subsystems or the overall system itself.

Fault Tolerance: The built-in ability of a system to provide continued correct operation in the presence of a specified number of faults or failures.

Fault Tree Analysis: A system safety analysis technique used as an inductive method (top down) to evaluate fault or failure events.

Fixed Crane: A crane whose principal structure is mounted on a permanent or semipermanent foundation.

Flammable: Any substance that is easily ignited and burns, or has a rapid rate of flame spread. Capable of being ignited and burning. With respect to a fluid or gas, means susceptible to igniting readily or to exploding.

Flashback Arrestor: A mechanical device utilized on a vent of a flammable liquid or gas storage container to prevent flashback into the container, when a flammable or explosive mixture ignites outside the container.

Flow Diagram: A scaled graphic/pictorial representation of the layout and locations of activities or operations and the flow paths of materials between activities in a process.

Fly-Fix-Fly: A description of the early approach to system safety, with reference to the aviation industry, that focused upon an after-the-fact method of designing safe systems.

Frequency Distribution: The tabulation of data from the lowest to the highest, or highest to the lowest, along with the number of times each of the values was observed or occurred in the distribution.

Frequent: In terms of probability of hazard or mishap occurrence, a hazard or event likely to occur numerous times during the life of an item.

Gas Law: The thermodynamic law applied to a perfect gas that relates the pressure of the gas to its density and absolute temperature.

Gauge Pressure: The pressure with respect to atmospheric pressure, or above atmospheric pressure as indicated on the appropriate pressure gauge. The difference between two absolute pressures, one of which is usually atmospheric pressure.

Gaussian Distribution: Pertaining to or having the appearance of a normal distribution.

Gear Ratio: The number of revolutions a driving gear requires to turn a driven gear one revolution. For a pair of gears, the ratio is found by dividing the number of teeth on the driven gear by the number of teeth on the driving gear.

General Duty Clause: Refers to Section 5 (a)(1) of the Occupational Safety and Health Act of 1970 which states: *Each employer shall furnish to each of his employees employment and a place of employment which are free from recognized hazards that are causing or are likely to cause death or serious physical harm to his employees, and shall comply with occupational safety and health standards promulgated under this Act.*

General Duty Clause Violation: Under the Occupational Safety and Health Act, a violation of the general duty clause exists when OSHA can show that the hazard is a recognized hazard, the employer failed to render its workplace free from the recognized hazard, the occurrence of an accident or adverse health effect was reasonably foreseeable, the likely consequence of the incident (accident or adverse effect) was death or a form of serious physical harm, and there exists feasible means to correct the hazard.

Grade D Breathing Air: Breathing air which meets the specifications of the Compressed Gas Association (CGA) Commodity Specification for Grade D air. It must have between 19.5% and 23% oxygen content and must contain maximums of

5 mg/m^3 condensed hydrocarbons, 20 ppm carbon monoxide, and 1000 ppm carbon dioxide; and it must have no pronounced odor.

Greater Hazard Defense: A well-established Occupational Safety and Health Review Commission (OSHRC) doctrine that, on some occasions, allows employers to escape sanctions for violations of otherwise applicable safety regulations because the act of abating the violation would itself pose an even greater threat to the safety and health of their employees.

Guarded: Covered, shielded, fenced, enclosed, or otherwise protected by means of suitable covers, casings, barriers, rails, screens, mats, or platforms to remove the likelihood of approach to a point of danger or contact by persons or objects.

Guideline: A recommended practice or other nonbinding suggestion issued by an agency, without the force of law.

Hazard: A condition or situation which exists within the working environment capable of causing an unwanted release of energy resulting in physical harm, injury, and/or damage.

Hazard Analysis: The analysis of systems, processes, and/or procedures to determine potential hazards and recommended actions to eliminate or control those hazards.

Hazard and Operability Study (HAZOP): A formal, structured investigative system for examining potential deviations of operations from design conditions that could create process-operating problems and hazards.

Hazard Correction: The elimination or control of a workplace hazard in accord with the requirements of applicable federal or state statutes, regulations, or standards.

Hazard Probability: The likelihood that a condition or set of conditions will exist or occur in a given situation or operating environment.

Hazard Severity: A categorical description of hazard level or degree, based upon real or perceived potential for causing harm, injury, and/or damage caused by a given hazard condition.

Hazardous Condition: Circumstances which are causally related to an exposure to a hazardous material.

Hazardous Pressure Systems: Systems used to store and transfer hazardous fluids such as cryogens, flammables, combustibles, hypergols, etc.

Health Hazard: A property of a chemical, mixture of chemicals, physical stress, pathogen, or ergonomic factor for which there is statistically significant evidence, based on at least one test or study conducted in accordance with established scientific principles, that acute or chronic adverse health effects may occur among workers exposed to the agent.

Health Hazard Analysis: The Health Hazard Analysis (HHA) is an analysis technique for evaluating the human health aspects of a system's design. These aspects include considerations for ergonomics, noise, vibration, temperature, chemicals, hazardous materials, etc. The intent is to indentify human health hazards during design and eliminate them through design features. If health hazards cannot be eliminated, then protective measures must be used to reduce the associated risk to an acceptable level. Health hazards must be considered during manufacture, operation, test, maintenance, and disposal (Ericson 2005).

Hertz (Hz): A measure of frequency in cycles per second (cps). The standard radio equivalent of frequency in cycles per second of an electromagnetic wave. Kilohertz (kHz) is a frequency of one thousand cycles per second. Megahertz (MHz) is a frequency of one million cycles per second.

Histogram: A graphical representation of two or more amplitude measures using rectangular shapes along either a discrete or continuous dimension. More commonly referred to as a bar graph or bar chart.

Hoist Angle: An angle at which the load line is pulled during a hoisting operation.

Human Error: The end result of multiple factors which influence human performance in a given situation. An often overused causal factor finding which, by itself, is not entirely descriptive of a true accident cause. Human error is considered more a symptom than a cause. See also *Human Factor.*

Human Error Probability (HEP): A measure of the likelihood of occurrence of a human error under special conditions:

$$HEP = error\ count/number\ of\ possibilities$$

Human Factor: Any one of a number of underlying circumstances or conditions which directly or indirectly affect human performance. These include physical as well as psychological factors that can potentially lead a person to make an error in judgment or action (human error) resulting in an accident. *See also Ergonomics.*

Human Factors: A combination of those aspects which effect human performance (such as personal, physical, psychological, situational, etc.) that may or may not contribute to an accident, incident, or near-miss occurrence. See also *Ergonomics.*

Human Factors Analysis: A systematic study of those elements involving a human–machine interface or other situation with the intent of improving working conditions, operations, or an individual's well-being. Also referred to as *ergonomic analysis.*

Human Factors Engineering: A concerted effort or attempt to design products and/or systems in consideration of human performance and those aspects which act upon or influence the human element of system operation.

Human Reliability: An assessment of the probability that an individual or group will adequately perform a given task at the appropriate time.

Hydraulic: Operated by water or any other liquid under pressure, including all hazardous fluids as well as typical hydraulic fluids that are normally petroleum based.

Hypothetical Question: In a What-If analysis, a form of question framed in such a manner as to call for an opinion from an expert based on a series of assumptions.

Ignition: The introduction of some external spark, flame, or glowing object that initiates self-sustained combustion.

Immediately Dangerous to Life and Health (IDLH): The maximum level to which a healthy individual can be exposed to a chemical for 30 minutes and escape without suffering irreversible health effects or impairing symptoms.

Imminent Danger: Any conditions or practices in a place of employment which are such that danger exists which could reasonably be expected to cause death or serious physical harm immediately or before the imminence of such danger can be eliminated.

Imminent Hazard: A hazardous situation, condition, or circumstance, the nature of which poses a serious and imminent threat to human health or the environment. If actions are not taken to immediately correct or stop the hazard cause, the results could be catastrophic.

Improbable: In terms of probability of hazard or mishap occurrence, a hazard or event whose occurrence is so unlikely during the life of an item, it can be assumed that the hazard will not occur.

Incident: An occurrence, happening or energy transfer which results from either positive or negative influencing events and may be classified as an accident, mishap, near-miss, or neither, depending on the level and degree of the negative or positive outcome.

Indirect Cause: A contributing causal factor other than direct cause associated with an incident.

Industrial Engineering: That engineering discipline concerned with the design, development, installation, and improvement of integrated systems of people, materials, equipment, and energy in the industrial environment.

Inspect: To verify quality, integrity, and/or safety through testing, observation, or other processes.

Intangible Risk: A risk involving unwanted consequences which are primarily non-physical, such as public opinion, employee morale, etc., but may have adverse effects (SSDC-11).

Interface: A common boundary or point of connection between two or more parts of a system or between systems, whether physical or perceptual.

Job Analysis: An evaluation of job requirements through an evaluation of the duties and tasks, facilities and working conditions, and worker qualifications and responsibilities necessary to perform a job.

Job Demand: The combined physiological, sensory-perceptual, and psychological requirements for, or loads experienced by a worker, performing a particular job.

Job Design: The process of determining what the job content should be for a set of tasks, how the tasks should be organized, and what linkage should exist between jobs.

Job Hazard Analysis: See *Job Safety Analysis.*

Job Safety Analysis: A generalized examination of the tasks associated with the performance of a given job and an evaluation of the hazards associated with those tasks and the controls used to prevent or reduce exposure to those hazards.

Usually performed by the responsible supervisor for that job and used primarily in the training and orientation of new employees. Also known as *Job Hazard Analysis*.

Job Severity Index (JSI): A guideline for matching job design and employee placement such that an acceptable risk of injury potential is present.

Kelvin Scale: A temperature scale with zero degrees equal to the theoretical temperature at which all molecular motion ceases.

Key Event: One incident which is primarily responsible for the time, place, and severity of an accident or other significant happening.

Knowledge-Based Behavior: A cognitive operating mode in which the individual attempts to achieve a goal in a situation with no clearly preestablished rules.

Ladder Safety Device: Any device, other than a cage or well, designed to eliminate or reduce the possibility of accidental falls and which may incorporate such features as life belts, friction brakes, and sliding attachments.

Lessons Learned: A formal, documented account or report of both the positive and negative aspects of operational or task experience which is compiled after the conclusion of the task. Used generally to highlight those actions which should or should not be allowed to occur during any subsequent performance of like or similar tasks.

Life Cycle: A phased concept to explain the various stages of product or system progression consisting of the concept phase, design phase, production phase, operational phase, and disposal phase. In system safety, the product or system life cycle is often used to indicate the timing of certain types of analytical evaluations.

Life-Cycle Characteristic Curve: See *Bathtub Curve.*

Life-Cycle Cost: The total cost of an item over its useful life, including purchase, maintenance, and operations.

Light Duty: A work classification in which an individual is not permitted to do heavy lifting for health or other reasons.

Limit Stop: Any device or mechanism which prevents further movement of a control, door, drawer, or other object at a certain point when motion beyond that point might have undesirable consequences. May be accomplished by audible click or tactile sensation.

Line and Staff Organization: In the structure of an organization, those members that are directly accountable and responsible for the daily operations of the enterprise are considered *Line* management and have the authority to implement/change company policy and operating procedures. Those that serve as advisors to the Line and can only recommend changes are considered Staff management.

Load Limit: The maximum weight or stress which an individual, floor, vehicle, or other structure can safely support.

Logic Gate: As pertains to fault tree analysis (FTA) and/or the Management Oversight and Risk Tree (MORT), a symbol used to identify the association between events on a logic tree.

Loss: Anything which increases costs or reduces productivity and has any adverse effect to the organization or society resulting from either normal operations or unplanned events (SSDC-11).

Machine Guard: Any piece of equipment or device on a machine intended to reduce or eliminate the chance of injury through the use of that machine.

Main Event: See *Contributory Event.*

Maintainability: An expression of the ability of a given product or system to be maintained (with minimum maintenance and repair) and remain in intended service throughout the operational phase of the product life cycle.

Marginal Event: An occurrence, subsequent to the introduction of a hazard or set of hazards into a system, that results in a level of injury, damage, or loss of minimal consequences. Quick recovery would be possible and probable. The parameters for this categorization are usually established by management in the System Safety Program Plan or other policy-making documentation.

Material Safety Data Sheet (MSDS): A compilation of data required under OSHA's Hazard Communication Standard on the identity of hazardous chemicals, their health and physical hazards, exposure limits, and precautions.

Maximum Expected Operating Pressure (MEOP): The highest pressure that a pressure vessel, pressurized structure, or pressure component is expected to experience during its service life and retain its functionality, in association with its applicable operating environments. It includes the effect of temperature, pressure transients and oscillations, vehicle quasi-steady and dynamic accelerations, and relief valve operating variability.

Mean: In statistical analysis, the arithmetic average derived from the addition of all value points in the sample, divided by the total number of points in the sample.

Mean Deviation: The average of the absolute deviations of values in a distribution from the mean.

Median: In statistical analysis, that value point which is precisely in the center (i.e., half the value points fall below the median and half lie above the median).

Methods Study: A systematic examination of the techniques, factors, and resources involved in the component parts of one or more operations, with the intent of improving techniques and productivity, while reducing costs.

Mishap: An occurrence which results in injury, damage, or both.

Mode: In statistical analysis, the most common or most frequent value that appears during evaluation or observation of a sample population of values.

Moment: A statistic measure, represented by the sum of the deviations from the mean, raised to some power, and divided by the number of terms used in accumulating the sum. Also, the tendency of a force to generate rotation in a body or torsion about an origin.

Monotony: The psychological state created by the lack of variety due to the repeated performance.

Near-Miss: An occurrence which had the potential to result in serious injury, damage, or both, but did not.

Negligible Event: An occurrence, subsequent to the introduction of a hazard or set of hazards into a system, that results in a level of injury, damage, or loss of such insignificant consequence that quick or total recovery would be highly probable and possible. The parameters for this categorization are usually established by management in the System Safety Program Plan, or other policy-making documentation.

Nip Point: The nearest point of intersection or near contact of two oppositely rotating circular surfaces or a rotating circular surface and a planar surface.

Node: Term used to identify individual system components when conducting a HAZOP study.

NonCausal Association: A statistical association in which no cause-and-effect relationship is apparent between two variables.

Nonrandom Sample: Any sample taken in such a manner that some members of the defined population are more likely to be sampled than others.

Normal Distribution: In statistical analysis, that distribution of events which occurs most often and is typically represented graphically as a bell-shaped curve.

Normal Event: As pertains to fault tree analysis (FTA) and/or the Management Oversight and Risk Tree (MORT), an event which occurs as a normal function in system operation that may or may not present a risk of hazard to that system. Represented graphically by a house shape in FTA and a scroll shape in MORT.

Occasional: In terms of probability of hazard or mishap occurrence, a hazard or event likely to occur sometime during the life of an item.

Operating and Support Hazard Analysis: A system safety analytical technique (also know as the *operational hazard analysis*) which focuses primarily on the hazards associated with or caused/enhanced by the human/task interface of system operations.

Operating Life: The period of time in which prime power is applied to electrical or electronic components without maintenance or rework.

Operations Process Chart: An abbreviated flow process chart consisting of a graphic/symbolic description providing a top-level view of the sequence for an entire operation, specifying such information as the actions and inspections involved, materials used, and pints of introduction, etc.

Oxygen-Deficient Atmosphere: An atmosphere which contains less than the approximately 20–21% of oxygen found in normal air; or, an atmosphere containing less than 19.5% oxygen by volume. It is that concentration of oxygen by volume below which atmosphere-supplying respiratory protection must be provided.

Oxygen-Enriched Atmosphere: An atmosphere containing more than 23.5% oxygen by volume.

Parameter: A characteristic of a population, such as the mean, standard deviation, or the variance. A variable quantity or arbitrary constant appearing in a mathematical expression, each value of which restricts or determines the form of the expression. Also, an arbitrarily defined constant value under a given set of circumstances and from which other values or functions may be defined.

Population: The total group of individual persons, objects, or items from which samples may be taken to estimate characteristics of that population by statistical methods.

Preassigned Probability: When the likelihood of all possible outcomes of a given event is known or can be determined, the probability of such outcomes are said to be preassigned (rolling dice, tossing a coin, etc.).

Preliminary Hazard Analysis (PHA): A system safety analysis method used to formally evaluate and document the hazard risks associated with a new or modified system.

Preliminary Hazard List (PHL): A "first-look" method of identifying potential or existing hazards associated with system design.

Primary Event: See *top event.*

Probability: An event that can reasonably be expected to occur on the basis of available evidence. The value of the ratio of the number of ways one or more specified events can occur to the total number of events which may occur. Expressed as a number between 0 and 1. The likelihood of observing a particular result or event, especially within a specified time or a given set of circumstances.

Probability Theory: In failure analysis, the examination of the likelihood of a specific failure or fault event, given a single opportunity for occurrence of that event.

Probable: In terms of probability of hazard or mishap occurrence, a hazard or event likely to occur several times during the life of an item.

Process Hazard Analysis: A thorough, orderly, and systematic approach to identify, evaluate, and control highly hazardous chemical processes. It involves a review of what could go wrong and what steps may be taken to safeguard against highly hazardous chemical releases.

Production Flow Analysis: The study of the routing of a part, component, or system through the various machines and workplaces and the operations it undergoes in a manufacturing or integration facility.

Program Manager: In military and/or other government agencies, as well as their contracting organizations, the term used to identify that person responsible for total contract management and administration, including the system safety effort. Likened to the vice president or general manager in the private sector.

Project Evaluation Tree: A system safety analytical technique which was developed from the more extensive management oversight and risk tree (MORT) method of analysis. A simplified and efficient method to evaluate a project or operation. Especially useful in the analysis of accidents and hazards.

Proximate Cause: The cause factor which directly produces the effect without the intervention of any other cause. The cause nearest to the effect in time and space.

Qualitative: The characteristic attributes or qualities pertaining to an exposure based on subjective information, nonrigorous quantitative data, and judgment.

Qualitative Risk Assessment: An examination of system risk based upon established criteria that allow the analyst to evaluate risk levels in relation to other risks or total system risk.

Quantitative: The property of anything which can be determined by measurement and expressed as a quantity.

Quantitative Risk Assessment: An application of statistical techniques to mathematically identify the level of probable risk associated with a given hazard, as it relates to total system operation.

Questioning Technique: A method for analyzing and attempting to improve work processes, generally by asking questions such as: (a) What is the purpose for some activity? (b) Why is a particular sequence followed? (c) Why does a particular person perform that job? and (d) Is the method being used to accomplish the task the best possible?

Reliability: An expression of the level of confidence that a given system or product will function, and continue to function, as intended throughout the life cycle.

Reliability Curve: See *Bathtub Curve.*

Remote: In terms of probability of hazard or mishap occurrence, a hazard or event whose occurrence during the life of an item is considered unlikely, but still possible.

Residual Risk: That risk which remains after the application or implementation of controls, barriers, or other risk-reducing methods or techniques.

Risk: The likelihood or possibility of hazard consequences in terms of severity and probability (Stephenson 1991). The probability of occurrence of a loss-producing event, the chance of loss. The probability or a range of probabilities that a specific adverse effect may occur under the conditions of human exposure. It may be expressed in quantitative terms, taking values from zero (certainty that harm will not occur) to one (certainty that it will). In many cases, risk can only be described qualitatively (i.e., as high, low, or trivial).

Risk Analysis: A detailed examination of any activity or functioning system in which potential adverse effects and their probabilities are calculated, and the various risks are quantified or measured (SSDC-11).

Risk Assessment: The qualitative and quantitative evaluation performed in an effort to define the risk posed to human health and/or the environment by the presence or potential presence and/or use of specific pollutants.

Risk Assessment Code: An alphanumeric rating of hazard risk based upon its anticipated frequency of occurrence and the resultant severity of exposure to such risk.

Risk Cost-Benefit Analysis: A combination of cost-benefit analysis and risk assessment. It is intended to assess the costs and benefits associated with prevention or reduction of risks to human health and the environment.

Risk Evaluation: An appraisal of the degree of undesirability of the various risks after they have been quantified. Consideration is given to the various factors and tradeoffs influencing risk acceptability (SSDC-11).

Risk Event: An occurrence with the potential to lead to an unwanted event such as an accident or incident.

Risk Factor: A correlation of characteristics (e.g., sex, age, race, obesity) or variables (e.g., smoking, occupational exposure level) with increased probability of a toxic effect.

Risk Management: The process, derived through the application of system safety principles, whereby management decisions are made concerning control and minimization of hazards and acceptance of residual risks (SSDC-11).

Root-Cause Analysis: With regard to compliance, an analysis which looks beyond superficial symptoms or underlying factors contributing to or causing shortcomings or failures in the system. It looks at something that occurred and asks what could have been done to have prevented it from happening in the first place.

Safe: A condition or situation that is free from hazards to health. Relatively free from the risk of danger, injury, or damage.

Safety: A measure of the degree of freedom from risk or conditions that can cause death, physical harm, or equipment/property damage (Leveson 1986).

Safety Critical: Any condition, event, operation, process, equipment, or system with a potential for major injury or damage (MIL-STD-1574A).

Safety Engineering: Discipline concerned with the planning, development, implementation, maintenance, and evaluation of the safety aspects of equipment, the environment, procedures, operations, and systems to achieve effective protection of people and property.

Safety Factor: The ratio of design burst pressure over the maximum allowable working pressure (MAWP) or design pressure; it can also be expressed as the ratio of tensile or yield strength over the maximum allowable stress of the material.

Safety Professional: An individual who, by virtue of specialized knowledge, skill, and educational accomplishments, has achieved professional status in the safety field (ASSE).

Safety Relief Valve: A valve fitted on a pressure vessel, or other containment under pressure, to relieve overpressure.

Safety Standard: Those standards designed to protect employees from hazards such as slips, trips and falls, lacerations and amputation from using machinery, fire hazards, and so on.

Sample Parameters: Estimators of population parameters such as the mean, standard deviation, etc. and are based on observations of a subset of the population.

Sneak Circuit Analysis: A system safety analytical technique (also known as *sneak analysis*) used to identify and evaluate the different possible ways in which inherent system design characteristics can either permit an undesired function to occur, prevent a desired function from occurring, or adversely effect critical operational timing. Typically associated with analysis of electrical or electronic systems and other energy transfer systems (pneumatic, hydraulic, etc.).

Software Hazard Analysis: A system safety analytical technique whose function is to evaluate potential faults in both operating system and applications software requirements, codes, and programs as they may effect overall system operation.

Soft Tree: Also known as Software Fault Tree Analysis, a system safety technique used to evaluate a single loss event and/or the effect of simultaneous failures with a software system on that single loss, or "top" event.

Standard Deviation: In statistical analysis, a value equal to the square of the variance.

Standard Error of the Mean: A measure of the variability of the distribution of sample arithmetic means with respect to the theoretical population standard deviation.

Statistical Analysis: A mathematical evaluation of past performance. In failure analysis, a focus on the total possible number of times a failure or fault event will occur given many opportunities for that occurrence.

Statistical Significance: An inference that the probability is low that the observed difference in quantities being evaluated could be due to variability in the data rather than an actual difference in the quantities. The inference that an observed difference is statistically significant is typically based on a test to reject one hypothesis and accept another.

Statistics: The field of applied mathematics which is concerned with the analysis, presentation, and derivation of conclusions from data.

Subsystem: An element of a system that, in and of itself, may constitute a system (MIL-STD-882).

System: A combination of people, procedures, facility, and/or equipment all functioning within a given or specified working environment to accomplish a specific task or set of tasks (Stephenson 1991).

System Critical: A single-point failure item or component in the system the loss or failure of which would result in a loss or failure of the entire system.

System/Subsystem Hazard Analysis: A system safety analytical technique used to evaluate hazards occurring on the subsystem or component level and the effect of their occurrence on overall system operations.

System Loss: Damage to an extent that renders repair impractical. Requires salvage or system replacement (MIL-STD-1574A).

System Safety: A subdiscipline of systems engineering that applies scientific, engineering, and management principles to ensure adequate safety, the timely identification of hazard risk, and initiation of actions to prevent or control those hazards throughout the life cycle and within the constraints of operational effectiveness, time, and cost (Stephenson 1991). The use of system engineering principles to provide a specified level of safety given the trade-offs involving cost, time, and the operations involved.

System Safety Analysis: A detailed, systematic method of evaluating the risk of hazard associated with a given system, product, or program. It utilizes a variety of techniques and approaches to accurately identify, resolve, or control exposure to those hazards.

System Safety Engineer: An engineer who is qualified by training, certification, and/or experience to perform system safety engineering tasks (MIL-STD-882).

System Safety Engineering: An engineering discipline requiring specialized professional knowledge and skills in applying scientific and engineering principles, criteria, and techniques to identify and eliminate hazards, or reduce the risk associated with hazards (MIL-STD-882).

System Safety Management: An element of management that defines the system safety program requirements and ensures the planning, implementation, and accomplishment of system safety tasks and activities consistent with the overall organizational requirements (MIL-STD-882).

System Safety Precedence: An ordered listing of preferred methods of eliminating or controlling hazards. Typically, it is listed as:

1. Design for minimum risk,
2. Incorporate safety devices,
3. Provide warning devices,
4. Develop procedures and training,
5. Acceptance of residual/remaining risk.

System Safety Program: The combination of tasks and activities of system safety management and system safety engineering that enhance operational effectiveness by satisfying the system safety requirements in a timely, cost-effective manner throughout all phases of the system life cycle (MIL-STD-882).

System Safety Program Objective: To reduce the risk of a given hazard or set of hazards to its lowest possible level of acceptance (as determined by management) without significant sacrifice of system effectiveness, operating schedules, or cost.

System Safety Program Plan: A written description of the planned method of implementing a system safety program in a given organization. It identifies responsibilities, objectives, system safety tasks to be performed, and the method of integrating the program into the organization's overall activities.

System Safety Tasks: Those activities, such as hazard analysis, associated with the system safety engineering discipline that are performed to accomplish the system safety program objective.

Task Analysis: An expansion of the Job Safety Analysis (JSA) method of identifying hazards associated with a given job or task. Differs from the JSA in its level of specific detail and consideration of the human interface in all aspects of the job performance.

Time-Loss Analysis: A specialized system safety analytical technique used to evaluate responses to accidents in consideration of the actual moment in time the response occurred following the accident. An evaluation is made of these responses and a determination of their effectiveness is made based upon losses that occurred up to the moment of intervention.

Top Event: As pertains to fault tree analysis (FTA) and/or the Management Oversight and Risk Tree (MORT), the primary fault event under analysis. Represented graphically as a rectangle.

Trial and Error: Pertaining to a blind, initially random, uninformed search for the correct solution or a path to that solution.

Undeveloped Event: As pertains to fault tree analysis (FTA) and/or the Management Oversight and Risk Tree (MORT), an identified fault event that will not be developed further because its occurrence has been determined insignificant with regard to its effect on the top event, or insufficient data exist to further evaluate the event, or the event is too complex for the purpose of a specific evaluation. Represented graphically by a diamond shape.

Unsafe Act: Any act or action, either planned or unplanned, which has the potential to result in an undesired outcome or loss (injury, property damage, lost production time, etc.). Conduct that causes an unnecessary exposure to a hazard or a violation of a commonly accepted procedure which directly permitted or resulted in a near-miss or the occurrence of an accident.

Unsafe Condition: Any existing or possible condition which, if allowed to continue, could result in an undesired outcome or loss (injury, property damage, lost production time, etc.). Any physical state that deviates from the accepted, normal, or correct practice and that has the potential to produce injury, excessive exposure to a health hazard, or property damage.

Useful Life: That period of time in the existence of a machine or system following any run-in phase and prior to the wear-out phase in which it is generally functionally stable in its operation.

Value Analysis: A systematic study to determine costs in each production phase for manufacturing an item, either during the engineering phase of product development or on an already existing product, generally with the intent to reduce costs by eliminating unnecessary steps.

Variance: A mathematical measure of the variation in the observed values of a sample population.

Vulnerability Analysis: Assessment of elements in the community that are susceptible to damage should a release of hazardous materials occur.

Wear-Out Phase: That period of time occurring after a system has performed much of its useful life and components begin to fail due to aging or other factors.

What-If Analysis: An informal but somewhat structured investigative method for introducing and evaluating hypothetical events, or series of events, associated with the operation of a given facility or process.

Work Environment: The physical location, equipment, materials processed or used, and the kinds of operations performed in the course of an employee's work, whether on or off the employer's premise, comprise the employee's work environment.

Work System: An integrated group of one or more machines and/or workers for coordinated activities in the output of some product or service.

Yield Strength: The stress at which a material exhibits a specified permanent deformation or set.

Zero-Fault Tolerant: Having no redundancy. Pertaining to a condition in which a single fault in a system will cause that system or the function performed by it to fail.

Zonal Analysis: A relatively new system safety analysis technique concerned with evaluating the geographic arrangement of installed systems, and its interconnections, as well as the influence of external events on those systems.

Bibliography

Abendroth, G. H. and Grass, J. M. 1987. A contracting program manager's guide to system safety. *Hazard Prevention* 4:14–19.

ANSI A17.1–1991. Elevators and Escalators. 1991. New York: American National Standards Institute, Inc.

Blackmane, H. S., Gertman, D. I., and Haney, L. N. September 1985. *The Process of Task Analysis (SSDC-31)*. U. S. Department of Energy, System Safety Development Center: EG&G Idaho, Inc., Idaho Falls, ID.

Briscoe, G. J. September 1982. *Risk Management Guide (SSDC-11R1)*. U. S. Department of Energy, System Safety Development Center: EG&G Idaho, Inc. Idaho Falls, ID.

Brown, D. B. 1976. *System Analysis and Design for Safety*. Englewood Cliffs, NJ: Prentice Hall.

Browning, R. L. 1980. *The Loss Rate Concept in Safety Engineering*. New York: Marcel Dekker.

Ericson, Clifton A. 2005. *Hazard Analysis Techniques for System Safety*. John Wiley and Sons.

Gloss, D. S., and M. G. Wardel. 1984. *Introduction to Safety Engineering*. New York: John Wiley and Sons.

Hammer, W. 1972. *Handbook of System and Product Safety*. Englewood Cliffs, NJ: Prentice Hall.

Johnson, W. G. 1973. *MORT, The Management Oversight Risk Tree*. Washington, DC: U.S. Atomic Energy Commission.

Knox, N. W. and Eicher, R. W. May 1983. *MORT User's Manual (SSDC-4R2)*. U. S. Department of Energy, System Safety Development Center: EG&G Idaho, Inc. Idaho Falls, ID.

Larson, M. S. and Hann, S. 1989. *Safety and Reliability in System Design.* Needham Heights, MA: Ginn Press.

Larson, M. S. and Hann, S. 1990. *Product Liability and Design Safety.* Des Plaines, IL: American Society of Safety Engineers (seminar handout manual).

Leveson, N. G. June 1986. Software safety: why, what, and how? *Computing Surveys* 18(2):125–163.

Leveson, N. "White Paper on Approaches to Safety Engineering." April 23, 2005.

Malasky, S. W. 1982. *System Safety Technology and Application.* New York: Garland STPM Press.

McCormick, E. J. 1976. *Human Factors Engineering and Design.* New York: McGraw Hill.

Roland, H. E. and Moriarty, B. 1983. *System Safety Engineering and Management.* New York: John Wiley and Sons.

Nertney, R. J. and M. G. Bullock. February 1976. *Human Factors in Design (SSDC-2).* U. S. Department of Energy, System Safety Development Center: EG&G Idaho, Inc. Idaho Falls, ID.

Nolan, D. P. 1994. *Application of HAZOP and What-If Safety Reviews to the Petroleum, Petrochemical, and Chemical Industries.* New Jersey: Noyles Publications.

NRI MORT User's Manual, 2nd edn, 2009 (ISBN 978-90-77284-08-7). Noordwijk Risk Initiative, in Partnership with the Royal Dutch Navy, Koninklijke Marine, The Netherlands.

Olson, Richard. E. Undated. *System Safety Handbook for the Acquisition Manager.* Air Force Space Division, Directorate of Safety. Distributed by the System Safety Society, Sterling, Virginia.

Spurr, W. A. and Bonini, C. P. 1973. *Statistical Analysis for Business Decisions.* Homewood, IL: Richard D. Irwin, Inc.

Stephenson, J. 1991. *System Safety 2000.* New York: Van Nostrand Reinhold.

Technical Analysis Incorporated (TAI). 1989. *System Safety Engineering* [Course Manual]. Houston, TX: TAI.

U.S. Air Force Regulation (AFR) 127-4. 1990. *Investigating and Reporting U.S. Air Force Mishaps.* Department of the Air Force, U.S. Government Printing Office, Washington, DC.

U.S. Department of Defense. August 1979. *MIL-STD-1574A: System Safety Program for Space and Missile Systems.* Department of the Air Force, U.S. Government Printing Office, Washington, DC.

U.S. Department of Defense. March 1984 (updated by Notice 1, 1987). *MIL-STD-882 1984: System Safety Program Requirements.* U.S. Government Printing Office, Washington, DC.

U.S. Department of Labor. July 1990. *Regulations Relating to Labor—General Industry.* Occupational Safety and Health Administration, Code of Federal Regulations, Title 29, Part 1910. Office of the Federal Register, National Archives and Records Administration, Washington, DC.

Index

Basic Guide to System Safety, Third Edition. Jeffrey W. Vincoli.
© 2014 John Wiley & Sons, Inc. Published 2014 by John Wiley & Sons, Inc.